Modeling the Radiation of Modern Sound Reinforcement Systems in High Resolution

Von der Fakultät für Elektrotechnik und Informationstechnik

der Rheinisch-Westfälischen Technischen Hochschule Aachen

zur Erlangung des akademischen Grades eines

Doktors der Naturwissenschaften genehmigte Dissertation

vorgelegt von

Diplom-Physiker

Stefan Feistel

aus Rostock

Berichter:

Universitätsprofessor Dr. rer. nat. Michael Vorländer

Universitätsprofessor Dr.-Ing. Dirk Heberling

Tag der mündlichen Prüfung: 22. Juli 2013

Diese Dissertation ist auf den Internetseiten der Hochschulbibliothek online verfügbar.

Stefan Feistel

Modeling the Radiation of Modern Sound Reinforcement Systems in High Resolution

Logos Verlag Berlin GmbH

λογος

Aachener Beiträge zur Technischen Akustik

Editor:
Prof. Dr. rer. nat. Michael Vorländer
Institute of Technical Acoustics
RWTH Aachen University
52056 Aachen
www.akustik.rwth-aachen.de

Bibliographic information published by the Deutsche Nationalbibliothek

The Deutsche Nationalbibliothek lists this publication in the Deutsche Nationalbibliografie; detailed bibliographic data are available in the Internet at http://dnb.d-nb.de .

D 82 (Diss. RWTH Aachen University, 2014)

ISBN 978-3-8325-3710-4
ISSN 1866-3052
Vol. 19

Logos Verlag Berlin GmbH
Comeniushof, Gubener Str. 47,
D-10243 Berlin
Tel.:   +49 (0)30 / 42 85 10 90
Fax: +49 (0)30 / 42 85 10 92
http://www.logos-verlag.de

Abstract

Starting from physical theory, a contemporary novel framework is developed for the acoustic simulation of sound radiation by loudspeakers and sound reinforcement systems. A variety of own measurements is presented. These agree very well with the predictions of the computer model considering practical uncertainty requirements.

First, small sound sources are discussed. Such a source is used at receive distances much larger than its characteristic dimension. A theoretical foundation is derived for the accurate reproduction of simple and multi-way loudspeakers using an advanced point source model that incorporates phase data. After that the practical implementation of this so-called CDPS model is presented including measurement requirements and the newly developed GLL loudspeaker data format. Subsequently the novel model is validated in detail by means of a number of different measurement results.

In the second part, larger systems are analyzed such as line arrays where the receiver is often located in the near field of the source. It is shown that theoretically any line source can be decomposed into smaller elements with a directional characteristic. This approach allows modeling the performance of the complete line source in both near and far field as long as the considered receive location is in the far field of the elementary sources. Several comparisons of measured line arrays display good agreement with predicted behavior and underline the superiority of this model compared to existing simulation methods. At the end theoretical methods and measured results are used to show for the first time that the influence of production variation among supposedly identical cabinets has a measurable but small effect on the overall performance of a line array.

The last part of this work deals with the consequences of fluctuating environmental conditions, such as wind and temperature, on the propagation of sound. In the context of this work, it is of particular interest to consider the coherent superposition of signals from multiple sources at the receive location. For this purpose a novel theoretical model is developed that allows predicting the mean variation of the propagation delay of the sound wave as a function of the statistical properties of the environmental parameters. Measurements of these properties as well as of the sound travel time are consistent with corresponding modeling results. Finally, it is discussed how the average total sound pressure level of a line array or loudspeaker arrangement is affected by the random variation of propagation delays.

A part of this work was distinguished with the AES Publications Award 2010. Parts of the proposed data format have been incorporated into the international AES56 standard.

Contents

Contents

1. Introduction

*Wie ein Singen kommt und geht in Gassen
und sich nähert und sich wieder scheut,
flügelschlagend, manchmal fast zu fassen
und dann wieder weit hinausgestreut -*

Rilke

Generating and perceiving sounds is natural for humans. This made communication with acoustic signals one of the pillar stones in the evolution of mankind. Even though our visual senses can process much more input and are used constantly throughout the day, our hearing system was crucial for the development of means that allow us to exchange objective information and subjective emotions. This is true for the crying baby, whose mother can immediately feel the level of urgency and the kind of need, as well as for the captivating lecture about highly abstract particle theory, the moving recitation of a sad poem written by Goethe, or the exhilarating sequence of rock songs in a concert of the Rolling Stones.

The development of human speech and music went hand in hand with the forming of increasingly complex social structures. In fact, in a social context the right mix of emotional and informational content in an acoustic signal can give exceptional power. For thousands of years, pathetic speech and the sound of war drums put soldiers in the right mood to risk their very life for an abstract cause. Huge cathedrals prove their "godly" nature by impressing visitors with the enveloping sound of preacher, organ and choir. And until today, leading politicians gain a large part of their authority by speaking to their audience on a personal and rational level combined.

These few examples show that the acoustic transmission of information plays an important role in our daily life. This role was expanded considerably by the invention of electricity which complemented the given acoustics of indoor rooms and outdoor environments by a completely new aspect, namely the ability to reproduce natural sounds with loudspeakers - at a higher volume, over a wider area and at a different point of time. The combination of both aspects, room or architectural acoustics on the one side and electro-acoustics or sound reinforcement on the other side, along with the increasing knowledge in the theory of sound waves gave rise to a variety of new acoustic concepts.

Nowadays we can roughly distinguish between three different fields of application where acoustic characteristics are of extraordinary importance. Venues such as concert halls, theaters, or lecture halls are built primarily for music and speech performances. They are optimized for an excellent acoustic impression. In stadiums, convention centers, or airports, so-called public address systems provide contextual information in a continuous manner. In addition to that, all public spaces support mass notification and voice evacuation systems, e.g. in malls, office buildings, or hospitals.

In particular since the events of September 11, 2002, more and more attention is given to speech intelligibility in occupied venues and the ability to direct warning messages locally. Other trends have also increased the awareness of acoustic circumstances. This includes the fact that modern architecture prefers buildings made of concrete, steel and glass. These materials usually cause substantial difficulties when a satisfying acoustic quality has to be established. Also, noise pollution by road traffic or air planes as well as health regulations for sound levels at music concerts have become topics that are commonly discussed and widely understood. Finally, the abundance of inexpensive computing power, may it be PC-based or DSP-based, has led to significantly advanced loudspeaker control concepts, such as digitally steered loudspeaker arrays.

With the increasing number of challenges and the growing awareness of the related solutions, more professional and reliable sound systems as well as appropriate room acoustic conditions are demanded. Gradually, design requirements have become more complicated, just like the available loudspeaker technology. For this reason, risk management had to be integrated step by step into the planning work, especially where measurable budgets are involved, e.g. for the installation of a sound system in a new building or for the adequate configuration of the sound system after the renovation of an existing building.

In this respect, modeling a sound system or a room before it is actually built offers numerous advantages. It allows looking at different scenarios at much lower cost than in practice. Simulating the properties of the acoustic system also allows reducing the overall costs by optimizing the type of equipment and the quantity of materials needed. Additionally, the model shared between the client, the design engineer, and the installer becomes the basis for better communication about the project. But modeling is not easy; building physical scale models is a complex and laborious task, and defining and solving mathematical models of acoustic systems manually is difficult. Only the advent of modern computers allowed acoustic modeling to become a widely employed tool in the design process.

Computer-based acoustic modeling software is usually fed with input data that describes the venue. After that a simulation run is performed which delivers results in the form of objective quantities that e.g. describe speech intelligibility over the audience areas. The results can also be evaluated subjectively by a process called auralization. Combining the acoustic characteristics of the simulated room and sound system with music or speech material recorded in a studio or another approximately anechoic room, one can listen to the sound in the virtual venue.

However, the quality of the modeling results relies substantially on the given input data and on the accuracy of the implemented simulation algorithms. For the user of the software it is therefore necessary to have detailed knowledge about the geometry of the room and the setup, about the acoustic properties of the surfaces of the room, and about the radiation characteristics of the acoustic sources. Information about the 3D geometry and the used surface materials can often be acquired from architectural drawings and material catalogues, respectively. But data for the loudspeakers are not

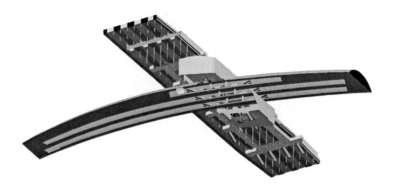

Figure 1.1.: Computer model of the main railway station of Berlin, Germany. (Courtesy ADA Acoustic Design Ahnert)

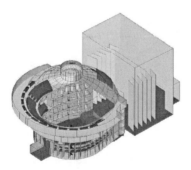

Figure 1.2.: Computer model of the Petruzzelli Theater in Bari, Italy. (Courtesy Daniele Ponteggia)

readily available; they must either be supplied by the manufacturing company or by a laboratory.

Unfortunately, even today there is only a rough, informal standardization in the pro-audio industry with respect to the type, quantity, and accuracy of loudspeaker data that are provided. On the other hand, the PC technology advancements of the last decade have eliminated many obstacles that related for example to the required measurement equipment, the processing of raw measurement data, the storing of finished data, and the resulting performance of the modeling software. This development has simplified the task of formulating conditions for modeling data and satisfying them in reality.

When establishing requirements for loudspeaker data and the related modeling algorithms, three primary aspects have to be considered for simulating the performance of modern sound systems:

1. The sound wave radiated by a single source must be reproduced in the model. A loudspeaker is typically described by sensitivity, maximum input or output level, and directional data, assuming a linear, time-invariant model. Care must be taken with respect to spectral and spatial resolution of the data as well as the delay and phase information included.

2. The combined sound field of several sound sources must be reproduced properly, both for multiple transducers in a box and for multiple boxes in an array or cluster.

3. The effects of the transmission medium on the propagation of the individual sound waves as well as on their superposition at the receive location have to be accounted for.

Only when these conditions are fulfilled, the computer model will be able to predict the direct sound field generated by the sound system.

In addition to that, any practical solution must also take into account the high degree of system configurability that is nowadays available, may it be with respect to e.g. the mechanical configuration of a flown line array that consists of many individual boxes with different splay angles, or to the electronic configuration that is determined by the delay, gain and filter settings of the DSP controller.

This work summarizes the results of several years of research and practical implementation in the direction of the above points. It focuses on the theoretical requirements for loudspeaker data as well as on the simulation algorithms related directly to sound sources.

In the following sections of this chapter a more detailed overview will be given over the different aspects of acoustic simulation. After that the main three chapters of this thesis will discuss the modeling of individual loudspeakers, of arrays of loudspeakers, and of their combined sound field at the receiver, respectively. The work will conclude with a summary and an outlook.

1. Introduction

1.1. Theoretical Background

As a starting point for the overview of acoustic simulation a few important theoretical definitions and relationships (see e.g. [1], [2], [3]) should be summarized. Mathematically, the propagation of sound is described by the linear *wave equation* for the sound pressure p as a function of space $\vec{r}$ and time t:

$$\frac{\partial^2 p}{\partial t^2} - c^2 \triangle p = 0, \tag{1.1}$$

where c is the speed of sound. Of primary interest are harmonic solutions with frequency ω of the form $p(\vec{r}, t) = p'(\vec{r})e^{i\omega t}$. These are determined by the *Helmholtz equation*

$$\triangle p + k^2 p = 0, \tag{1.2}$$

where $k = \omega/c$ is the wave number. An important solution of this wave equation is the outgoing *spherical wave*

$$p(\vec{r}, t) = \frac{A}{|\vec{r} - \vec{r}_0|}e^{i(\omega t - k|\vec{r} - \vec{r}_0|)}, \tag{1.3}$$

where A is the complex amplitude and $\vec{r}_0$ the location of the wave's origin. Equation (1.3) is a solution of the homogeneous wave equation for $\vec{r} \neq \vec{r}_0$. But it also solves the inhomogeneous wave equation that contains a source term with periodic driving at location $\vec{r}_0$ and amplitude A. Because of this the relationship (1.3) is often referred to as the radiation behavior of an omnidirectional *point source*.

The *superposition* principle for sound waves stems from the fact that linear combinations of solutions of the wave equation are solutions of the wave equation, as well. Consequently, the sound field generated by a set of point sources j can be described by

$$p(\vec{r}, t) = \sum_j p_j(\vec{r}, t) = \sum_j \frac{A_j}{|\vec{r} - \vec{r}_j|}e^{i(\omega t - k|\vec{r} - \vec{r}_j|)}. \tag{1.4}$$

The so-called *Eikonal equation* [1], [4] is an important derivative of the wave equation. It assumes solutions of the form $p(\vec{r}, t) = A(\vec{r})e^{i\phi(\vec{r}, t)}$, where $\phi(\vec{r}, t) = \omega(t - \Theta(\vec{r})/c_0)$ is the phase, c_0 a reference speed, and $\Theta(\vec{r})$ is termed the Eikonal. In this picture, local wave fronts are established by surfaces of constant $\Theta(\vec{r})$. Assuming wavelengths that are small compared to other characteristic dimensions of the system, a condition for the Eikonal can be derived as a function of the local speed of sound $c(\vec{r})$,

$$(\nabla\Theta)^2 = \frac{c_0^2}{c^2}. \tag{1.5}$$

The Eikonal equation (1.5) represents the high-frequency limit of the wave equation. It lays the foundation for *ray-tracing* methods and *particle*-based considerations regarding the propagation of sound waves. Besides several other results of this equation, the travel

5

time Δt along the ray path s can be directly calculated by integrating the local speed of sound over the distance of interest:

$$\Delta t = \int\limits_0^s \frac{1}{c(\vec{r})} ds, \qquad (1.6)$$

where $\vec{r}$ must be expressed as a function of s.

1.2. Numerical Computation

This section will give a brief look at the current state of the art of electro-acoustic and room acoustic modeling [5], [6]. Naturally, such a consideration will not be able to cover the diversity of acoustic modeling approaches and their numerical foundation in detail. It will also not be concerned with the theory and the terminology of room acoustics, because it is of no further relevance to the investigations related to modeling sound sources. However, this outline should provide a practical context for the following chapters and it also represents a part of the motivation for this research work. Especially Sections 1.2.2 and 1.2.3 will give an introduction into the scope of this thesis.

The first section of this part, 1.2.1, will be concerned with principal computational models whereas the subsequent sections will discuss essential problems related to the numerical implementation of the boundary conditions, primarily with respect to acoustic modeling based on ray-tracing.

1.2.1. Acoustic Simulation

A main goal of acoustic simulation in electro-acoustic and room acoustic applications is to find the solution of the Helmholtz equation (1.2), when subjected to given boundary conditions. Although the wave equation itself seems to be fairly simple, the difficulty of the task arises from the complexity of formulating and treating complicated boundary conditions properly. Only for very simple enclosures, such as the famous shoebox model, the particular solution of the wave equation obeying the boundary conditions can be given analytically. For most other bounded volumes numerical methods must be applied. Obviously, the inclusion of sound sources and of sound attenuation by the propagation medium further increase the complexity.

In practice, the solution of the wave equation is usually sought in the form of the system's complex-valued, time-independent transfer function, for a given source or a set of sources and a receive location. More specifically, the impulse response in the time domain and its counterpart, the frequency response in the frequency domain, provide the basis for a number of quantities that have been empirically found to establish useful criteria for different aspects of the acoustic quality of an environment.

One distinguishes between two fundamentally different numerical methods. On the one hand, the so-called high-frequency or particle-based solutions have been developed based on the Eikonal theory (1.5), very similar to the development in the field of optics.

They have found wide usage but are limited by the assumptions of the Eikonal approximation and can thus not be applied without restriction, in particular, when considering wavelengths that are long compared to the characteristic dimensions of the room.

On the other hand, so-called low-frequency or wave-based approaches numerically solve the wave equation itself. Although this method is by definition more accurate and principally applicable to the entire frequency range, its numerical complexity, the necessary computation times, as well as the difficulty of defining the boundary conditions in a practical manner allow for only limited application, mainly in modeling of small rooms at low frequencies.

A commonly accepted dividing line between these two fields of application is given by the Schroeder frequency.

Schroeder Frequency and Reverberation Onset

In one of his many works, Schroeder[1] derived a spectral limit which separates the frequency range of clearly distinguishable room modes from the frequency range of strongly overlapping room modes. This frequency depends primarily on the volume of the room and on its reverberation time [7], [8].

Because of its simplicity this approximation has found widespread use and is applied until today when speaking of a room's modal part of the spectrum and its diffuse part of the spectrum, or of wave acoustics versus geometrical acoustics. It is also used to roughly determine the lower frequency limit of a particle-based or the upper limit of a wave-based solution that is appropriate for a given room.

Similar to the spectral domain, a line can be drawn in the time domain as well. When considering sources in a room with reflecting surfaces and thus a reverberant field, a prominent point of time in the impulse response is the reverberation onset. At that time the diffuse, homogeneous, and isotropic field of reverberant sound is considered as approximately established and discrete, early reflections cannot be isolated anymore. An estimate for this point of time is given by [9]. It is based on mean-free-path assumptions and relates to the room volume as well as to the center time of the first-order reflections.

Mean-Free-Path Statistics

The simplest and commonly practiced approach of estimating the transfer function of a room, at least in parts, is the so-called mean-free-path statistics [9], [10]. It usually involves the computation of the direct sound arrivals at the receiver utilizing the point source solution given in (1.3) and the estimation of the energy of the reverberant field by employing the room's surface area, its volume, and the average sound absorption coefficient of the surface materials, according to Sabine's or Eyring's formulas, for example. The energies contained in the direct sound field and in the reverberant sound field are

---

[1]The author was very honored by having met Mr. Schroeder, who laid the scientific foundation for many applications in acoustics, in excellent shape at the 2006 ASA meeting in Honolulu. Unfortunately Mr. Schroeder passed away in 2010 at an age of 89.

computed on the basis of the acoustic output level and the directional radiation characteristics of the sources. By this means the distribution of received sound energy over time can be approximated for any location in the room, and in consequence a number of fundamental measures can be derived.

Because this approach assumes by definition that the reverberant field is homogeneous and isotropic and that it decays exponentially after disabling the source of sound energy, no specific statements can be made about discrete reflection patterns, such as echoes, or e.g. the dominant direction of received reverberant energy at a certain point of time. Being based on the Eikonal concept, this statistical solution cannot account for any wave-based effects, either.

Ray-Tracing Methods

Going beyond the statistical estimate of the reverberant field there exist a variety of different approaches based on the particle- or ray-like propagation of the sound wave, see e.g. [5], [11], [12], [13], [14], [15], [16], [17]. They all have in common that reflection paths are computed by taking into account the geometry of the room boundaries relative to the source and the receiver. Usually one or several of the following methods are employed:

- A systematic search for important reflections by "scanning" the room's surface. This may be accomplished by sending out rays from the source using a defined angular resolution or by directly computing possible reflection paths utilizing the mirror-image model. Because of the deterministic nature of this method, the rays arriving at the receiver are known by their exact arrival time, frequency spectrum, direction, as well as the particular sequence of reflective surfaces they have passed.

- A Monte Carlo approximation of the specular reflections of the reverberant field by radiating sound particles from a source in random directions and tracing them until a receiver is hit. Each particle carries information about its energy content, travel time, angle of incidence, and possibly its phase. Because this method is based on issuing a large number of particles from each sound source, some information on the particle's history usually has to be discarded for restricted computer memory.

- A Monte Carlo approximation of the diffuse part of the reverberant field by generating non-specular or scattered reflections at the boundaries and tracing them to the receiver.

In order to determine the room's transfer function most accurately all three strategies have to be combined in a hybrid approach [18]. That is because the first method is computationally expensive and cannot provide scattered reflections. Depending on the number of particles, the second method using random numbers may not find all significant reflections, although its results must converge with the first method for very high particle numbers and long calculation times. The third method is needed in order to establish the "energy floor" of the reverberant field, that is, the diffuse reflections that represent the main part of the late reverberation. Approaches not implementing this last but computationally expensive step have often resorted to estimating artificially a so-called random tail from extrapolating the early reflections.

Over the last decade a number of extensions have been proposed in order to incorporate first-order wave-based effects into the ray-tracing model [19]. This includes in particular the diffraction of sound waves by edges. Besides such effects ray-tracing methods often suffer from two other, mostly more significant sources of error, namely an inaccurate description of the radiation properties of the sound source (see Section 1.2.2) and a lack of accuracy of the simulated reflection process that until today employs absorption coefficients measured in the diffuse field, i.e. independent of the angle of incidence, as well as mostly estimated scattering coefficients (see Section 1.2.5).

Wave-Based Methods

Wave-based methods always require the discretization of space, either as a mesh of the boundary surface, such as in the Boundary Element Method (BEM, [20], [21]), or as a mesh of the volume itself, such as in the Finite Element Method (FEM, [22], [23]) and in the Finite-Difference Time-Domain method (FDTD, [24]). Additionally, also a discretization in the frequency domain (BEM, FEM) or in the time domain (FDTD) is necessary.

After that, the BEM and FEM approaches result in a system of linear equations (or an eigenvalue problem for the eigenmodes) whose solution involves very large matrices. The size of the matrix and thus computational effort and memory requirements are directly related to the resolution of the spatial mesh. The typical choice of a sixth or an eighth of a wavelength for the mesh resolution represents an upper limit of the frequency range, for example, 500 Hz for a studio room modeled on a contemporary PC [25].

For the modeled range of time or frequency a complex-valued transfer function is obtained which can be used for any further investigation.

Similar to the high-frequency solution, also for the low-frequency solution the major uncertainties stem from the incomplete knowledge about the acoustic properties of the boundary in practice, that is, the complex input impedance of the walls, and from the imprecise model of the radiation behavior of the sources and their interaction with the room. This fact limits the broad application of the approach in practice until today.

Results

The primary result of acoustic simulation is the transfer function of the room, one data set for each particular combination of sources and receive location. Especially ray-tracing models may also provide directional information for the received direct sound and individual reflections[2].

---

[2]Wave-based methods can supply these data as well, but in a less immediate way.

In a post-processing step these data are then evaluated in two different manners:

- Calculation of objective quantities, that is, acoustic criteria, such as sound pressure level, reverberation time, or speech intelligibility, which are mostly related to the temporal structure of the arriving sound energy [26], [27], [28], [29], [30].

- Subjective evaluation of the acoustic impression by means of auralization, which is the process of making audible the transfer function or impulse response, or its convolution product with dry audio material [5], [31], [32].

Supported by the increasing performance of modern computers, recently the aforementioned low-frequency solutions have become more viable options, especially from a technical point of view. This has led to a number of theoretical investigations regarding the potential convergence and merge of the wave-based calculation results and their particle-based counterparts. Although nowadays an overlap in the frequency range of the calculated transfer functions can be achieved computationally, the two methods do not converge necessarily, due to their very different approaches and their different treatments of the boundary conditions. Therefore appropriate cross-over algorithms have to be found which allow establishing a broadband transfer function without introducing new artifacts. First research results indicate that regular crossover filters as known from loudspeaker design can provide satisfying answers [33].

1.2.2. Sources

In acoustic modeling one distinguishes primarily between three types of sound sources:

1. Electro-acoustic sources, such as professional loudspeakers, are approximately linear transducers converting an electric input signal into an acoustic output signal. They usually have a fixed location and orientation in the room and radiate sound energy with certain spectral, temporal, and directional characteristics [1], [28], [34].

2. Natural sources, such as speakers, singers, or instruments, radiate an acoustic output signal. They are typically modelled with a certain output level, frequency spectrum, and directional radiation behavior [35], [31].

3. Small noise sources, such as machines or air vents, and large noise sources, such as airplanes or highways, generate an acoustic output signal. Depending on the type of source the radiation can be angle-dependent. Noise sources usually have a characteristic spectrum and output level [36], [37].

A large part of acoustic modeling is concerned with the first type of sources. That is because of the commercial need to design expensive loudspeaker systems properly but also because of the relative simplicity of measuring the acoustic performance parameters of a loudspeaker and incorporating them in a computer model. The modeling of the second type of sources represents an important task in applications where no or little sound

(a) Wireframe drawing.   (b) 3D radiation pattern at 2 kHz, "balloon" plot.

Figure 1.3.: Computer model of a loudspeaker line array.

reinforcement is being used. Although various natural sources have been measured and successfully modeled in room acoustic applications, their overall (commercial) relevance is comparably small. The third type of sources plays a role, on the one hand when the signal-to-noise ratio relative to an electro-acoustic or natural source has to be considered, or on the other hand when the emission and the propagation of acoustic noise is critical. Some noise sources such as heavy machinery are difficult to measure with respect to their acoustic radiation properties and are therefore usually characterized only by their sound power output.

This work will be concerned mainly with appropriate modeling of electro-acoustic sources, although some of the results can be applied to other types of sources, as well. Despite the fact that the accurate representation of the loudspeaker in acoustic simulation, especially in the domain of particle-based models, was the matter of many discussions, commonly used implementations lacked precision, generality, or practical feasibility, as will be shown beginning in Section 1.3.

The insufficiencies of the first generation of loudspeaker data sets for modeling purposes became especially obvious over the course of the last decade when mechanical line arrays [38], [39], digitally steered loudspeaker columns [40], [41], DSP-controlled loudspeakers [42], [43], [44], and other configurable, highly flexible, and accordingly complex sound sources [45] were increasingly employed. Simultaneously, the advancing availability of computing power and memory for the acoustic modeling process made step by step obsolete any artificial data reduction and restrictive assumptions about the loudspeaker's performance [46], [47], [48], [49], [50].

The most important parameters for modeling the sound radiated by an electro-acoustic source are the sensitivity of the loudspeaker, its maximum output regarding sound pressure level, and the angle dependence of its transfer function [51]. The idea of representing the transfer function of the loudspeaker in an angle-dependent way is principally based on the particle-model, eq. (1.5), and assumes that the wave front at the receiver is approximately plane, both during the data acquisition as well as in the simulation. This assumption allows the simulation process to treat the loudspeaker like an omnidirectional point source with a directional correction. In the model the source can thus be considered as a source of rays whose initial energies are determined by the output level and by the directional correction for their respective emission angle.

Generally, loudspeaker data are discretized spectrally, that is, in the frequency domain, as well as spatially, that is, with respect to angle[3]. The required and practically feasible spectral and angular resolutions as well as the necessity of a real-valued or complex-valued transfer function are important aspects regarding the achievable accuracy of the reproduction of the radiated sound field in the computer model. These issues will be addressed in Chapters 2 and 3 of this thesis.

Mainly in applications of loudspeaker design and optimization the radiation of a loudspeaker is often simulated utilizing a BEM computer model which allows investigating wave-based effects, like the interaction between the transducer and the housing [52], [53], [54], [55]. While such a model is time-consuming both with respect to building it and with respect to running calculations, it can be very accurate. Because of the high computational effort it is normally not used directly in combination with a room acoustic model, but it can be utilized to supply directional transfer functions instead of measuring them. This is particularly useful for large loudspeaker systems which are impractical to be measured as a whole. This auxiliary method was made use of in Chapter 3.

In both FEM and BEM approaches, sound sources can be implemented using appropriate source terms in the formulation of the wave equation or by incorporating adequate boundary conditions [56], [57].

1.2.3. Medium

Air being the propagation medium of interest, there are two significant effects that are normally accounted for in particle-based acoustic models. On the one hand, the dependence of the speed of sound on the environmental parameters such as air temperature [1] must be taken into account because the propagation speed is essential to determine, for example, delay times between loudspeakers. On the other hand, the attenuation of the propagating sound wave in air is significant in the high-frequency range and over long distances. The absorption of sound energy due to friction losses in the modulated carrier also depends on environmental parameters, such as the temperature and humidity of air as well as the absolute air pressure [58].

In addition to considering environmental parameters when describing the propagation medium, a number of research works also discuss the feasibility of including solid objects

---

[3]A continuous representation of loudspeaker data by means of a functional description and a number of coefficients in either domain is possible but has so far not proven to be more efficient or more accurate in practice.

of given density and approximately homogeneous and isotropic spatial distribution, such as trees in a forest [59] or people in an arena [60], in the definition of the properties of the medium.

Whereas the above effects can still be modeled by assuming a homogeneous and isotropic medium, other factors of importance require more effort, such as stable vertical temperature gradients in indoor and outdoor applications or steady air drafts (indoors) and wind (outdoors). Although these effects have not yet found much consideration in commercial packages for indoor acoustic modeling, they are accounted for in outdoor noise propagation simulations as well as in related academic research [61], [62], and in practical installations [63], [64].

Another condition for most acoustic models is the time invariance of the medium. Obviously, the appropriate simulation of a medium changing over time requires a discussion of time scales. If the changes happen very slowly compared to the characteristic dynamics of the sound field, one can easily assume that the system is time-invariant for any simulated point of time. If the changes happen on time scales similar to the propagation time of sound, the model itself will have to include the changes which can become very complicated. Changes on very fast time scales can be ignored in the modeling process in the sense that an average state of the medium can be assumed as long as the fluctuations are relatively small.

Compared to deterministic temporal changes of the propagation medium, random fluctuations, such as turbulence due to wind gusts or air convection, are even more difficult to treat [65]. This type of problem requires a stochastic approach. This topic will be discussed in detail in Chapter 4.

1.2.4. Receivers

With respect to sound receivers, one can distinguish between two different types:

1. Electro-acoustic receivers, such as microphones, are transducers that convert an acoustic input signal into an electric output signal. They are described by a distinct sensitivity, directionality, as well as an amplitude range of sound pressure where they work effectively [29], [34].

2. Natural receivers, primarily the human ears, convert an acoustic input signal into an electric output signal which is directly fed to the brain for processing [29]. For the purpose of room acoustic modeling, each ear can be described by its sensitivity as well as by its directional transfer function [5]. In some applications the hearing threshold as well as masking due to high pressure levels play a role [27].

Although there have been many discussions during the last years about standardizing microphone specifications and measurements, for describing such receivers there has not yet evolved a fully accepted standard on which acoustic modeling could be based [66], [67]. On the other hand, most microphones exhibit one out of a few typical directional patterns, such as an omnidirectional or cardioid behavior. One of the main concerns until today is the correct measurement and reproduction of the proximity effect. It occurs when the sound source is relatively close to the microphone and the wave front of the incident sound wave cannot be considered as flat anymore [68], [69].

From this perspective, the potential distance dependence of the received sound relative to the source has been largely neglected in room acoustic modeling. Most microphone data sets are provided for the far field. Typically these consist of a set of magnitude-only frequency responses for different angles and are based on axial symmetry.

For the purpose of auralization (see Section 1.2.1), the measurement of the receive properties of the human ears is crucial. In addition to the frequency-dependent sensitivity, the outer part of the left and of the right ear are each described by a set of angle-dependent head-related transfer functions (HRTF). These are measured by placing a small measurement microphone at the entrance of the ear canal of a human or of a model of the human head and taking measurements for all spatial directions [70], [71], [72].

In contrast to the limited availability of data sets for professional microphones, HRTF data sets are almost abundant despite the greater measurement complexity. HRTF data are usually provided either for an average, representative human head, a so-called dummy head, or for the head of a particular person. Since the ultimate goal is a satisfying binaural auralization the head-related transfer functions have to be prepared carefully using complex data with sufficient spectral and angular resolution [5], [73].

Given the impulse response with level, time, spectral, and directional information for each arriving pulse as the result of a particle-based simulation, the binaural impulse response (BIR) of the modeled human head in a room can be immediately computed by convolving each pulse with the HRTF that corresponds to its angle of incidence at the receiver. Once the BIR is defined, binaural auralizations can be performed by convolving the BIR with a dry signal, such as speech or music recordings, and listening to the result, preferably with a headphone. A cross-talk cancelation setup (XTC) can be used for binaural auralization, as well [74], [75].

A similar procedure is used for the auralization of results in listening rooms. The B-format and other data formats [76], [77] capture directional information comparable to the BIR but for a larger number of different directions. These channels are later on combined in a synthesis process where a specific mix of these directional data is computed as a filter for each of the available playback loudspeakers. Depending on the algorithm the optimal listening zone in the room may be limited to a so-called sweet spot or allow for a larger use area.

In fact, a lot of research revolves around the exact reproduction of the human listening experience from auralizing modeling results compared to the physical impression on-site. This also includes the use of head-tracking systems for better interaction with the simulated sound field [78] as well as complete virtual-reality environments where the listener can even walk through a simulated room [79], [80].

1.2.5. Boundaries

In order to model the sound field inside a room it is necessary to define the boundary conditions, that is, the geometrical layout and the acoustical properties of the room's surfaces. In particle-based modeling approaches it is usually most efficient to subdivide the surface of the room into triangles in order to obtain plane and geometrically simple elementary objects. This process is also called tesselation. Systematic errors may be introduced if non-planar surface elements, such as curved walls, are not treated carefully [5]. It should also be noted that, in principle, the precision of the geometrical model should be adapted to the acoustical relevance of the modeled surfaces as well as to the accuracy of the drawings of the building. In reality, the knob of a door will not contribute any significant reflections to the room response at any location and one cannot be sure about its exact size and placement relative to the room, either. In practice a degree of detail up to a resolution of approximately 10 cm has proven useful.

In addition to the location and size of a wall, also its acoustic properties have to be determined. In room acoustic simulations based on particle models, primarily the diffuse-field absorption coefficient is being used in order to account for the attenuation of the traveling sound wave when it is reflected by a boundary. There are a number of issues connected with that. The most important ones are:

- Having been measured in a reverberation chamber, the diffuse-field absorption coefficient of a surface material does not provide any directional information [81]. This may lead to high uncertainties in the model regarding the estimated level of a first-order reflection, for example [82], [83].

- The diffuse-field absorption coefficient does not provide any information about the phase change of the reflected wave relative to the incident wave [84]. This is especially relevant for the first floor reflection and glancing effects [81], [83].

- The common measurement method is based on the comparison of the reverberation time of the reverberation chamber with and without the material sample [85]. This procedure may cause systematic errors, e.g. due to edge diffraction by the boundaries of the sample [9]. In specific cases, this method may even yield an "absorption coefficient" greater than 1.

In addition to these limitations acoustic modeling approaches normally assume the surface to be locally reacting. This means that the reflection of a sound wave at one point of the surface does not cause the emission of sound from other areas of the surface or affect the reflection of another sound wave at any other point, for example, by vibrations of the surface structure [82].

The second important acoustic property of the boundary is its scattering behavior. The scattering coefficient is defined as the ratio between energy that is reflected in a non-specular way and all reflected energy. Also here severe limitations apply because the diffuse-field scattering coefficient used in acoustic modeling usually lacks any directional information [86], [87], [88]. Furthermore, it is difficult or impossible to actually measure the scattering coefficient of certain wall structures, e.g. of steps or pillars. In such cases reasonable estimates must be made.

The scattering coefficient must not be confused with the diffusion coefficient. Whereas the first describes the part of energy reflected non-specularly, the latter defines the angular homogeneity of the reflected sound wave [82]. The diffusion coefficient is of less importance for acoustic modeling but it it considered a quality criterion for acoustic diffusers.

For low-frequency simulation, such as FEM, the knowledge about the phase of the reflected wave is also important as it defines the actual location of the nodes and of the extrema of a room mode. This means that the complex reflection factor or the input impedance of the wall must be known. There exists a wealth of published models for estimating the complex reflection factor as well as the complex transmission factor for single- and multi-layered walls [89], [36]. However, their physical input data, like the flow resistivity of a porous material is often not known in practice.

Even if these data are available, other important factors like the finite size of the surface element and its connection to other surface elements remain to be considered. Step by step, progress is made in this field [56], [22], [23], [90]. But compared to the modeling process in geometrical acoustics the practical applicability of the wave-based modeling approach is still more limited.

Both absorption coefficient and scattering coefficients are usually given as magnitude-only data in a resolution of 1/3rd or 1/1 octave bands. In the ray-tracing process the attenuation of the particle's energy when reflected at a surface is usually taken into account by reducing its energy in proportion to the absorption coefficient. The scattering coefficient can be used to determine the additional reduction of the particle's energy due to scattering losses when it is reflected specularly. It can also be used as a probability for the particle to be reflected non-specularly [12], [18].

1.3. Conventional Point Source Models in Electro-Acoustic Simulation

In the following sections the discussion of modeling data for electro-acoustic sound sources starting in Section 1.2.2 will be continued. Now the historical development of loudspeaker data in acoustic modeling with an emphasis on their accuracy will be reviewed and advantages and disadvantages will be outlined. This overview provides the formal and conceptual motivation for the research work presented in this thesis.

1.3.1. Directional Point Source Model

Based on the time-independent version of the propagation equation (1.3) for a point source one can define the sound pressure

$$p(\vec{r}, \omega) = \frac{A(\vec{r}/|\vec{r}|, \omega)}{|\vec{r}|} e^{-ik|\vec{r}|}, \tag{1.7}$$

where $A(\vec{r}/|\vec{r}|, \omega)$ represents the complex directional radiation function. It is construed of an absolute radiation amplitude $A'(\omega)$, a complex on-axis transfer function (or sensitivity) $\Gamma_0(\omega)$, and a complex directional transfer function $\Gamma(\vec{r}/|\vec{r}|, \omega)$ so that $A = A'\Gamma_0\Gamma$. For uniqueness of the separation, the directional transfer function Γ is by convention normalized to $\Gamma_0(\omega)$ for all frequencies in a distinct direction, such as the loudspeaker axis. It can be considered as a frequency- and angle-dependent correction factor for the omnidirectional point source (1.3). The factor Γ could be derived, for example, from a multi-pole expansion or a measurement in the far field.

Obviously, this approach makes implicit use of the Eikonal assumption discussed in Section 1.1. The wave front is radiated by the point source into each direction $\vec{r}/|\vec{r}|$ with a corresponding amplitude A. Each segment of this wave front propagates into the room along a straight line and expands like the surface section of a spherical wave whereas the sound pressure at any point of the wave front decreases according to the conservation of acoustic energy. Being in the far field of the source, the wave front can be considered as locally plane at any point of the propagation path [1].

From the perspective of a modeling algorithm that deals with loudspeakers of a sound reinforcement system, eq. (1.7) is a good starting point. In practice, any loudspeaker in the system is defined by its type, its location and orientation, as well as its drive voltage:

- The loudspeaker type provides the directional transfer functions Γ_0 and Γ that are obtained from measurements or simulation of the radiation properties of the enclosure.

- The drive voltage of each loudspeaker determines the corresponding radiation amplitude A'.

- Location and orientation of each loudspeaker define the individual transformation of the coordinate system of the loudspeaker type into the coordinate system of the room.

Assuming an isotropic, homogeneous medium and lossless propagation, eq. (1.7) can be used immediately to determine the direct sound pressure level of a source at any location in the room and for any frequency.

1.3.2. Discretization

Real-world loudspeakers exhibit a fairly complex radiation behavior both with respect to their frequency response as well as with regard to their directionality. As a result, the most practical way of describing a loudspeaker in a model is a discrete data set. That of course requires two distinct steps, namely a) discretizing the definition space of $A(\vec{r}/|\vec{r}|, \omega)$ and b) defining an appropriate interpolation function that re-establishes continuity in the modeling domain for frequencies and angles that are not covered by measured data points,

$$p(\vec{r}, \omega) = \frac{g(A(\theta_j, \phi_k, \omega_l), \vec{r}/|\vec{r}|, \omega)}{|\vec{r}|} e^{-ik|\vec{r}|}. \tag{1.8}$$

Here the spherical coordinates θ_j, ϕ_k and the frequencies ω_l are discrete data points in the space of $\vec{r}/|\vec{r}|$ and ω, respectively. The function g represents the interpolation function in the modeling domain. Since A is only determined at the given measurement points one can replace it formally and identically by a complex-valued, three-dimensional matrix $\hat{A}$:

$$p(\vec{r}, \omega) = \frac{g(\hat{A}_{j,k,l}, \vec{r}/|\vec{r}|, \omega)}{|\vec{r}|} e^{-ik|\vec{r}|}. \tag{1.9}$$

Obviously the discretization raises questions regarding the required angular and spectral resolution in order to minimize sampling errors [49], [91]. Indirectly related to that is the question of an appropriate interpolation scheme for complex data. This non-trivial problem will be discussed further in Chapter 2. The average uncertainty of the discretization and interpolation over frequency and angle can be estimated by

$$\delta A \sim \frac{1}{\int_\omega d\omega \int_\Omega d\Omega} \sqrt{\int_\omega d\omega \int_\Omega d\Omega \left| g(\hat{A}_{j,k,l}, \vec{r}/|\vec{r}|, \omega) - A(\vec{r}/|\vec{r}|, \omega) \right|^2}, \tag{1.10}$$

where Ω is the definition space of $\vec{r}/|\vec{r}|$.

Also the uncertainty of the measurement of $\hat{A}_{j,k,l}$ will affect the accuracy of the interpolation result. Unfortunately, information about the measurement accuracy is seldom published or not even acquired. As shown in Chapter 3.4, in practice this kind of uncertainty is related to the production spread of loudspeaker samples of the same type. If loudspeaker data sets are accompanied by information about the measurement uncertainty they should also provide information about sample-to-sample variation. Ideally, the acquisition of loudspeaker data would include the measurement of a number of different loudspeaker samples in order to estimate that effect.

1.3.3. Conventional Data Formats

For several decades, any loudspeaker - in its entirety - was considered a point source in acoustic modeling. Therefore a single, fixed data matrix like $\hat{A}$ was the preferred method of describing its directional radiation behavior, no matter how simple or complicated the loudspeaker or loudspeaker system was in reality. Several reasons led to that fact:

- For many years, the processing and memory capabilities of the computer platform where the calculations take place used to be very limited. In order to leave enough room for the simulation itself, loudspeaker data had to have small memory requirements and be easy to manage.

- For similar reasons, the available measurement equipment established significant constraints regarding performance and memory on the acquisition side. Also, early measurement platforms were not even able to measure complex-valued transfer functions or to apply windowing in order to post-process measurement data.

- The distribution of loudspeaker data from the measurement platform to the final user had to be relatively fast and simple.

Historically, the usage of tabular data formats did not only cause wide-spread debates about the necessary data resolution and adequate interpolation algorithms. But it also required discussing conversion and reduction methods for the raw measurement data in order to fit the given format[4] and it raised the problem of standardized data sets across different loudspeaker manufacturers. Table 1.1 shows a selection of different conventional data formats used by acoustic simulation programs [92], [93], [94], [95]. Clearly, the resolution increased gradually over time. But these predominantly assume that the loudspeaker can be represented sufficiently by a single point source with a fixed directional behavior.

Format	Frequency Resolution	Angular Resolution	Complex Data	Minimum Symmetry	Year of Introd.
EASE 1	1/1 Octave	15°	No	Quarter-Sphere	1990
EASE 2	1/1 Octave	10°	No	Half-Sphere	1994
ULYSSES 2.8 UNF	1/3 Octave	5°	No	Full-Sphere	1996
CADP2 GDF	1/3 Octave	5°	No	Full-Sphere	1996
EASE 3 SPK	1/3 Octave	5°	No	Full-Sphere	1999
EASE 4 SPK	1/3 Octave	5°	Yes	Full-Sphere	2002
CLF 1	1/1 Octave	10°	No	Full-Sphere	2005
CLF 2	1/3 Octave	5°	No	Full-Sphere	2005
CLF 2 v2	1/3 Octave	5°	Optional	Full-Sphere	2011

Table 1.1.: Commonly used tabular formats for loudspeaker data. The object-oriented GLL format [48] discussed in this work was first introduced in Dec 2005.

1.3.4. Compromises and Errors

During the last decade configurable loudspeaker systems have become widely used, including mechanically configured line arrays, digitally steered column loudspeakers, or DSP-controlled crossover systems (see Section 1.2.2). Obviously, changing the mechanical or electronic configuration of the loudspeaker system results in a different radiation behavior. This level of flexibility cannot be reproduced in the modeling domain when using a fixed data table.

In addition to this fundamental new requirement a number of other aspects were awaiting solutions:

---

[4]Often this process of transforming the measured data was accompanied by inaccuracies.

1. Introduction

- The assembly of multiple loudspeakers in a cluster requires a good reproduction of the interaction between the individual loudspeakers. Data formats without phase information suffer from a significantly increased uncertainty.

- Especially for crossover systems the so-called acoustic center, that is, the apparent origin of the spherical sound wave, changes its location with frequency. This poses a problem for the representation of loudspeakers by a magnitude-only data set with a fixed reference position if that position is used for interference calculations based on travel time.

- For practical reasons, loudspeaker data must also specify a maximum input voltage. This situation cannot be accounted for properly when systems with multiple inputs are to be treated like a single loudspeaker with one input.

- Even more so, the effect of internal filter settings on the maximum input voltage for each pass-band, of pass-bands relative to each other and relative to the input signal cannot be considered in a rigid, single-input model.

- Historically, the maximum input capability of loudspeaker systems is given by a power rating based on an almost artificial impedance rating, whereas actual testing is performed by determining maximum input voltages that are unambiguous.

- Beyond only providing the configuration options of the real loudspeaker, it would be beneficial if the end user could configure the loudspeaker in the model in a way similar to the real device instead of viewing and manipulating raw data tables.

- From a practical perspective, the management of loudspeaker data in two distinguished steps is desirable. As a first step, the loudspeaker company or loudspeaker designer assembles the raw data and compiles the data set into a fixed file. In a second step the end user can view the distributed loudspeaker data and manipulate them only in the way it was intended by the creator.

- A range of questions is concerned with the practical measurements required in order to model loudspeaker arrays, including measurement setup and resolution. Related topics are the influence of elemental variation on the array performance as well as the coherence between radiating array elements.

These issues will be addressed and largely resolved by the methodology proposed in this thesis. However, a number of issues that are also relevant for the accurate modeling of sound sources remain open. These concern primarily wave-based effects in combined loudspeaker systems that cannot be captured by the point source approach, such as the accurate treatment of diffraction and shadowing by the loudspeaker enclosure or by neighbor cabinets. Coupling effects between low-frequency transducers or between a transducer and the room are not considered, as well. A question that is also not addressed here is how natural sources and noise sources should be represented in the acoustic model.

2. Modeling Small Sound Sources

In the introductory chapter conventional loudspeaker data formats and their accuracy have been discussed. This chapter will be concerned with proposed improvements regarding the inclusion of phase information in the data set, the data resolution, as well as measuring conditions. Based on that, a new, more general concept for the representation of loudspeaker data for computational purposes will be introduced, namely the Generic Loudspeaker Library (GLL) format, an object-oriented description language. The chapter concludes with several exemplary, detailed comparisons of prediction results with measurements for a two-way loudspeaker, loudspeaker clusters, loudspeakers with filters, as well as column loudspeakers.

This chapter of the thesis will be concerned primarily with *small sound sources*. Here they are defined as electro-acoustic sources that are used mainly in their far field, that is, at receive distances that are large compared to the spatial extent of the sources. Accordingly, such sources can be assumed to act like point sources. In the next chapter *large sounds sources* will be treated. A large source is generally used into its near field and cannot be approximated by a single point source anymore.

2.1. Significance of Phase Data

In order to simulate the acoustic properties of combinations of sound sources on the basis of the point source model (1.9) it is crucial to include not only magnitude data but also phase data [96], [97]. A practical motivation for that will be given now and then a theoretical analysis of the problem will be provided.

When the acoustic performance of a sound system is modeled, the contribution of each loudspeaker or sound source to the overall sound pressure level is summed at the receiving location. This sum can be calculated as a power sum regardless of the relative phase of the sound arrivals or it can be calculated as the coherent sum of the sound waves by accounting for propagation time differences and for the inherent phase response of each source[1]. Right now, the correct treatment of the propagation time will be the primary focus because in comparison the inherent part is normally a slowly varying function over radiation angle and frequency. As it will be shown, the proposed approach accounts for both aspects well enough.

If the loudspeaker directional data consist only of magnitude information, the propagation phase is conventionally computed based on the distance between the loudspeaker

---

[1]A more realistic model will be introduced in Chapter 4 where coherence will be considered as a function of environmental parameters, frequency, and propagation distance.

and the receiver location in the room model. In order for the relative phase relationship between two sources in the model to be accurate, the loudspeaker locations in the model must correspond to the actual locations of the acoustic sources in reality. This actual location was also termed the so-called *acoustic center* of the loudspeaker (e.g. [98], [99]). Determining the acoustic center is a difficult problem in practice, for a number of reasons:

- Multi-way systems with several transducers active in the same frequency range cannot be represented by a single point.

- Crossover systems exhibit a frequency-dependent location of the acoustic center.

- The membrane of the transducer or of the wave guide is spatially extended and thus the location from which the sound wave effectively radiates is not always well defined.

These problems are usually addressed by choosing the center of the sound transducer that covers the highest frequency range as the acoustic center because this minimizes the error when combining loudspeakers with magnitude-only data based on propagation delay.

In order to overcome these issues and to increase the accuracy of the prediction, complex-valued data should preferably be used to describe a loudspeaker's directional behavior. For this purpose, phase data have to be acquired as part of the directional measurement process. They need to be compensated for the propagation delay from the loudspeaker location to the measurement point in order to be applied to other propagation distances in the later model.

It will be shown that the actual choice of the point of rotation used for the directional measurements is not relevant as long as it is in the vicinity of the active sound source(s) of the loudspeaker. That is because the acquired phase data will contain the information about the spatial offset of each source relative to the point of rotation. As a result, the insertion location of the loudspeaker in the room model is simply and unambiguously given by the point of rotation during the directional measurement. The definition of the acoustic center becomes irrelevant and no longer required.

Obviously, this entire consideration only applies if the measured data are to be combined with other data in a complex manner. A simple coverage calculation based on power summation does not require phase information and will thus not be subject to phase-based errors. Also, purely energy-based considerations that do not take into account the phase, often found for example in room acoustic ray-tracing, will not suffer from the described problem.

2.1.1. Overview

If the directional response of a loudspeaker is measured and only magnitude data are used, the frequency response of a sound source under a certain angle is given by $\hat{A}_{i,k,l} \equiv |\hat{A}_{i,k,l}|$, see eq. (1.9). This common model can be compared with a more precise approach, namely the use of complex data, where the matrix $\hat{A}_{i,k,l}$ is a complex-valued quantity. For this purpose the systematic errors that arise from the different treatment of the measured directional data will be estimated.

Deriving the data set $\hat{A}_{i,k,l}$ from measurements normally consists of two steps, namely a) measuring at a fixed, well-defined measurement distance and b) referencing the measurement data back to the point of rotation[2]. This process is imaged by the subsequent analysis in order to derive the associated uncertainties for each type of data [97].

Concept

The calculation is led by the following thoughts:

1. It is postulated that the loudspeaker to be evaluated is known in every detail. This *Exact Model* is assumed to be given and serves as a measure further on. It can also be understood as a virtual, ideal reference measurement.

2. Then restrictions are applied to the calculation of these data. The restrictions correspond to actual conditions in the real world and result in models that are different from the exact model, such as the *Magnitude Model* or the *Complex Model*. Naturally, the accordingly changed results will exhibit errors of unknown quantity.

3. Comparing the calculation results of these limited models with the exact model data will allow obtaining error limits and approximations.

Without loss of generality the following considerations are restricted to a purely vertical arrangement of one or several point sources and one or several points used as the reference locations for a measurement. Also, any practical issues and errors arising from the practical measuring conditions are not considered. In this section, errors are evaluated that are inherently made by using different abstract data models to represent an acoustic source.

Setup

To start with, it is assumed that a loudspeaker with a single acoustic source can be safely reduced to a point source. This is justified as long as all considerations are limited to the far field of the source [1]. The complex pressure p of the spherical wave radiated by

---

[2]In the following sections the terms *reference point* and *point of reference* as well as *POR* will be used as synonyms for *point of rotation*.

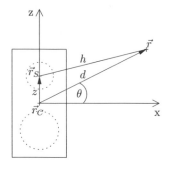

Figure 2.1.: Loudspeaker setup and coordinate system relative to point of rotation $\vec{r}_C$.

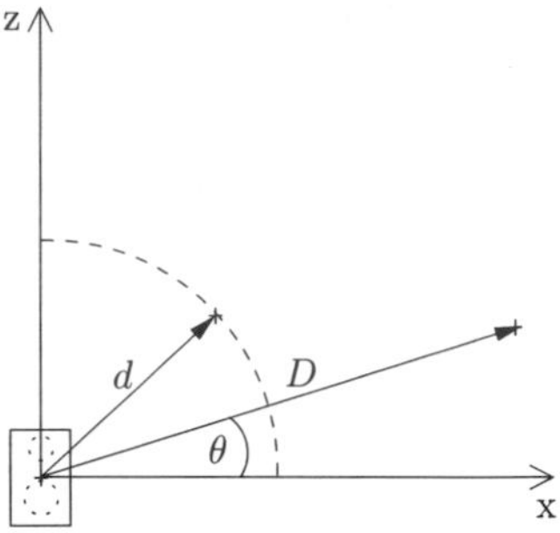

Figure 2.2.: Loudspeaker measurement at distance d and extrapolation to simulated receive location (D, θ).

the source located at $\vec{r}_S$ is accordingly:

$$p(\vec{r}, \omega) = \frac{A}{|\vec{r} - \vec{r}_S|} e^{-ik|\vec{r} - \vec{r}_S|}, \tag{2.1}$$

with the complex directional radiation factor $A = A(\frac{\vec{r} - \vec{r}_S}{|\vec{r} - \vec{r}_S|}, \omega)$ and the wave number $k = \frac{\omega}{c}$. The consideration may be restricted to two dimensions by defining the receive location $\vec{r}$ as a function of the measurement distance d and the angle θ relative to the origin $\vec{r}_C$:

$$\vec{r} = \vec{r}_C + d\cos\theta\vec{e}_x + d\sin\theta\vec{e}_z, \tag{2.2}$$

where $\vec{e}_x$ and $\vec{e}_z$ are the unity vectors in the x- and z-direction, respectively. The distance between source and measurement point is

$$h(d, \theta) = |\vec{r} - \vec{r}_S|. \tag{2.3}$$

The source location is given by

$$\vec{r}_S = \vec{r}_C + z\vec{e}_z, \tag{2.4}$$

where z represents the vertical offset of the acoustic source from the origin (Fig. 2.1).

The origin of the coordinate system, $\vec{r}_C$, is identical with the point of rotation for the later loudspeaker measurement. Expressed in new coordinates, the sound wave is defined by

$$p(d, \theta, \omega) = \frac{A(\theta, \omega)}{h(d, \theta)} e^{-i\frac{\omega}{c}h(d, \theta)}, \tag{2.5}$$

where it was assumed that A is a slowly varying function of the angle θ, and $d \gg z$, so that $A(\frac{\vec{r} - \vec{r}_S}{|\vec{r} - \vec{r}_S|}, \omega) \approx A(\theta, \omega)$.

Next, the post-processing steps applied to the measurement data in reality must be accounted for. Looking from the point of measurement and assuming that the wave front is radiated from the point of rotation, one can refer back by the measurement distance d and describe the amplitude of this virtual point source as

$$A_{Vir}(d, \theta, \omega) = d e^{i\frac{\omega}{c}d} p(d, \theta, \omega). \tag{2.6}$$

This yields the effective directional radiation function A_{Vir} based on the actual value of A:

$$A_{Vir}(d, \theta, \omega) = \frac{d A(\theta, \omega)}{h(d, \theta)} e^{-i\frac{\omega}{c}(h(d,\theta)-d)}. \tag{2.7}$$

This is the analytical expression for the apparent directional radiation function of a point source that is offset relative to its assumed origin. The propagation equation for this virtual point source in the model is thus:

$$p_{Vir}(D, \theta, \omega) = \frac{A_{Vir}(d, \theta, \omega)}{D} e^{-i\frac{\omega}{c}D}, \tag{2.8}$$

where D is the receive distance from the origin (Fig. 2.2). Finally the following expression is obtained for the sound pressure of a wave measured at fixed measurement distance d, referenced back and computed at receive distance D:

$$p_{Vir}(D, \theta, \omega) = \frac{d A(\theta, \omega)}{D h(d, \theta)} e^{-i\frac{\omega}{c}(D-d+h(d,\theta))}. \tag{2.9}$$

The systematic error caused by this treatment stems from the fact that a point-source like behavior is extrapolated from the measurement radius to any other radius relative to the reference point of the measurement.

Models

In the *Magnitude Model*, the measurement data are referenced back to the reference point, as in equation (2.8), but for the spherical wave front propagating from this point only the magnitude as a function of the angle is considered. This is equivalent to

$$p_{Mag}(D, \theta, \omega) = \frac{|A_{Vir}(d, \theta, \omega)|}{D} e^{-i\frac{\omega}{c}D}. \tag{2.10}$$

Inserting the effective radiation function (2.7) yields

$$p_{Mag}(D, \theta, \omega) = |A(\theta, \omega)| \frac{d}{D h(d, \theta)} e^{-i\frac{\omega}{c}D}. \tag{2.11}$$

Note that here the only term correcting for the spatial mismatch of source location and reference point is found in the magnitude factor $\frac{d}{h(d,\theta)}$.

In the *Complex Model*, the measurement data are referenced back to the reference point, as in eq. (2.8). But for the spherical wave front propagating from this point, both magnitude and phase data are considered as functions of the angle. This is given by

$$p_{Com}(D,\theta,\omega) = \frac{A_{Vir}(d,\theta,\omega)}{D}e^{-i\frac{\omega}{c}D}. \tag{2.12}$$

More precisely, this reads in terms of the actual radiation function:

$$p_{Com}(D,\theta,\omega) = A(\theta,\omega)\frac{d}{Dh(d,\theta)}e^{-i\frac{\omega}{c}(D-d+h(d,\theta))}. \tag{2.13}$$

There are two correction factors that account for the offset of the reference point: one for the amplitude, $\frac{d}{h(d,\theta)}$, and a second term for the phase, $h(d,\theta) - d$.

In the *Exact Model*, the wave front is calculated exactly; the directional measurement is basically performed at the receiver location, see eq. (2.5). This means

$$p_{Exact}(D,\theta,\omega) = p_{Vir}(D = d,\theta,\omega), \tag{2.14}$$

which equates to

$$p_{Exact}(D,\theta,\omega) = \frac{A(\theta,\omega)}{h(D,\theta)}e^{-i\frac{\omega}{c}h(D,\theta)}. \tag{2.15}$$

2.1.2. Analysis for Single Source

In order to quantify the errors associated with the Magnitude Model and the Complex Model both will now be compared with the Exact Model. The error for the magnitude of a single source can be defined by

$$\Delta A = \frac{|p_X(D,\theta,\omega)|}{|p_{Exact}(D,\theta,\omega)|}, \tag{2.16}$$

where p_X represents the pressure according to eqs. (2.11) or (2.13), respectively. Note that ΔA is defined as a multiplicative error that is smallest when its logarithm vanishes. The phase error can be defined in the following manner:

$$\Delta\phi = |\arg p_X(D,\theta,\omega) - \arg p_{Exact}(D,\theta,\omega)|. \tag{2.17}$$

Error Estimate for Magnitude Model

Eq. (2.16) can be evaluated using eqs. (2.11) and (2.15). Also, the measuring distance d and the receiver distance D are assumed to be large:

$$\frac{z^2}{d^2} - \frac{2z}{d}\sin\theta \ll 1, \qquad d \ll D. \tag{2.18}$$

Then a Taylor expansion can be performed of which only terms of first order are kept. A straightforward calculation yields:

$$\Delta A^{(Mag)} = 1 - \frac{z^2}{2d^2} + \frac{z}{d}\sin\theta. \tag{2.19}$$

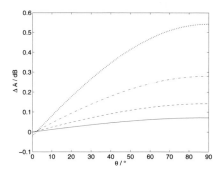

Figure 2.3.: Amplitude error of both Magnitude Model and Complex Model, $d = 3$ m, offset $z = 0.025$ m $(-)$, $z = 0.05$ m $(- -)$, $z = 0.1$ m $(-.-)$, $z = 0.2$ m $(\cdots)$.

The error for the amplitude is quite small. For a typical value of $z = 10$ cm, $d = 3$ m, and the worst case of $\theta = 90°$ it is about 0.3 dB. As shown in Fig. 2.3, the error increases with angle θ and offset z between source and reference point.

For the phase error one can make similar assumptions as in eq. (2.18). For simplicity, it is also assumed that the inherent phase of the source is zero, $\arg A = 0$. In the following discussion, this error would have to be added to the phase error of the Magnitude Model. Applying again a first order approximation by omitting terms of higher order one finds:

$$\Delta\phi^{(Mag)} = \frac{\omega}{c} \left| \frac{z^2}{2D} - z\sin\theta \right|. \tag{2.20}$$

For large distances, as $D \to \infty$, the error scales with the offset z and the angle θ (for θ not too small):

$$\Delta\phi^{(Mag)} = \frac{\omega}{c} z |\sin\theta|. \tag{2.21}$$

Figure 2.4 shows the dependence of the phase error on frequency f and angle θ. For increasing angles and increasing frequency the phase error grows. On the loudspeaker axis $(\theta = 0°)$ the error is negligible in first-order terms, whereas perpendicular to the system axis the error is maximal. Figure 2.5 shows the dependence of the phase error on frequency f and on the offset z between source and reference point. For increasing frequency and for increasing offset the error grows, as well.

Error Estimate for Complex Model

Obviously, the error of the amplitude is the same for the Magnitude Model and for the Complex Model,

$$\Delta A^{(Com)} = \Delta A^{(Mag)}. \tag{2.22}$$

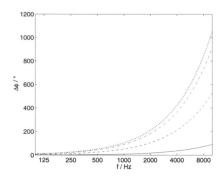

Figure 2.4.: Phase error of the Magnitude Model, $z = 0.1$ m, directions $\theta = 5°$ (—), $\theta = 30°$ (- -), $\theta = 60°$ (-.-), $\theta = 85°$ (· · ·).

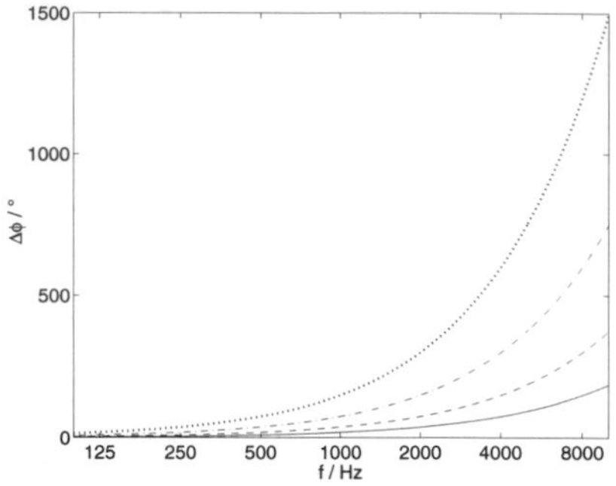

Figure 2.5.: Phase error of the Magnitude Model, $\theta = 45°$, offset $z = 0.025$ m (—), $z = 0.05$ m (- -), $z = 0.1$ m (-.-), $z = 0.2$ m (· · ·).

However, the error of the phase turns out to be quite different. The assumptions (2.18) are applied again. But this time the second order terms of the Taylor expansion need to be retained, because main and first order terms cancel out. As a result the error in second order terms is found:

$$\Delta\phi^{(Com)} = \frac{\omega}{c}\frac{z^2}{2}\left(\frac{1}{d} - \frac{1}{D}\right)\cos^2\theta. \tag{2.23}$$

This result yields the expected, namely that under an angle of $\theta = 90°$ the Complex Model is exact, because in the vertical direction the propagation phase differential due to the offset between the source and the point of reference is constant over distance.

For large distances, as $D \rightarrow \infty$, the error scales with the square of the offset z and it depends also on the angle θ:

$$\Delta\phi^{(Com)} = \frac{\omega}{c}\frac{z^2}{2d}\cos^2\theta. \tag{2.24}$$

Note that in contrast to the error in the Magnitude Model, here the error is greatest on the system axis ($\theta = 0°$). Furthermore comparing with the Magnitude Model, it can be concluded that the Complex Model reduces the error for the phase by at least an *order of magnitude*.

Figure 2.6 shows how the phase error depends on the frequency f and the angle θ. When increasing the frequency the error grows, but it becomes smaller when increasing the angle θ. In comparison to the Magnitude Model, the phase error of the Complex Model is almost a hundred times smaller. Figure 2.7 shows the frequency dependence of the phase error for various values of the offset z. For smaller distances between source and reference point the error decreases. Also here the error is much smaller than for the Magnitude Model, approximately forty times. The next Figure 2.8 depicts the

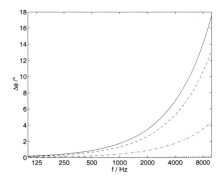

Figure 2.6.: Phase error of the Complex Model, $z = 0.1$ m, $d = 3$ m, directions $\theta = 5°$ (−), $\theta = 30°$ (- -), $\theta = 60°$ (-.-), $\theta = 85°$ (···).

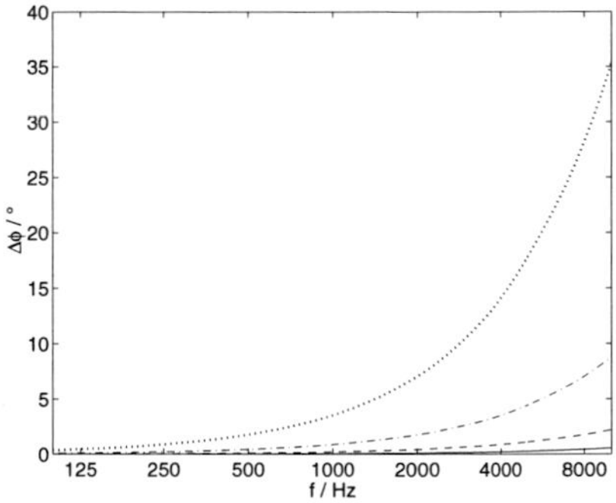

Figure 2.7.: Phase error of the Complex Model, $\theta = 45°$, $d = 3$ m, offset $z = 0.025$ m (−), $z = 0.05$ m (- -), $z = 0.1$ m (-.-), $z = 0.2$ m (···).

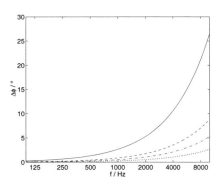

Figure 2.8.: Phase error of the Complex Model, $\theta = 45°$, $z = 0.1$ m, measuring distance $d = 1$ m (−), $d = 3$ m (- -), $d = 5$ m (-.-), $d = 10$ m (···).

dependence of the phase error on the measuring distance d. The phase error is smaller when the directional measurement is performed further away from the reference point.

Critical Frequency

In the previous sections an analytical expression was derived for the phase error that is caused by the offset of the point of rotation from the actual location of the source. In order to put this into a practical context one can define a critical limit that indicates when the phase error $\Delta\phi$ exceeds $\pi/4$ or $45°$. This choice of the maximally allowed phase error is somewhat arbitrary. Because the phase error grows monotonously with frequency the so-called *critical frequency* can be introduced. It represents the frequency at which the tolerable phase error is exceeded.

For the Magnitude Model, eq. (2.21) yields a condition for the upper frequency limit (for θ not too small):

$$f_{Crit}^{(Mag)} = \frac{c}{8z|\sin\theta|}. \tag{2.25}$$

In the typical case of $z = 10$ cm and $\theta = 45°$, the upper frequency limit would accordingly be $f_{Crit} \approx 600$ Hz.

With respect to the Complex Model, the result (2.24) for the phase error yields a condition for the upper frequency limit:

$$f_{Crit}^{(Com)} = \frac{cd}{4z^2 \cos^2\theta}. \tag{2.26}$$

For a typical setup of $z = 10$ cm and $d = 3$ m this results in an upper frequency limit of $f_{Crit} \approx 25.5$ kHz (at $\theta = 0°$). Vice versa, the error that is introduced at $f = 8$ kHz for such a setup is $\Delta\phi \approx 14°$. Although this error is quite small, care must be taken when two or more sources are combined. In particular in critical directions, where destructive interference leads to a steep slope of the pressure function, the resulting error of the pressure sum function may still be significant.

Figures 2.9 and 2.10 show the critical frequency for both models, eqs. (2.25) and (2.26), as a function of the offset z between source and reference point. These graphs are of much practical interest as they outline the error that is a-priori included in the two data models, Magnitude Model and Complex Model. Any of the two models should be used only below its critical frequency. For any offset z between source and reference point there is a critical frequency that represents the upper limit for applicability within a given error. It is obvious that the more accurate Complex Model allows for a much higher critical frequency than the Magnitude Model.

As a practical example, a two-way loudspeaker with a high-frequency unit (HF source) and a low-frequency unit (LF source) is considered. The two source points are spaced 20 cm apart. When the point of rotation is chosen exactly in the center, equal to $z = 0.1$ m, the Magnitude Model has an upper frequency limit of $f_{Crit} \approx 425$ Hz (choosing $\theta = 90°$ where the error is maximal). The limit is the same for both sources, LF and HF. This

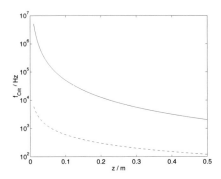

Figure 2.9.: Critical frequency of the Magnitude Model (- -) and Complex Model (−), $\theta = 45°$, $d = 3$ m.

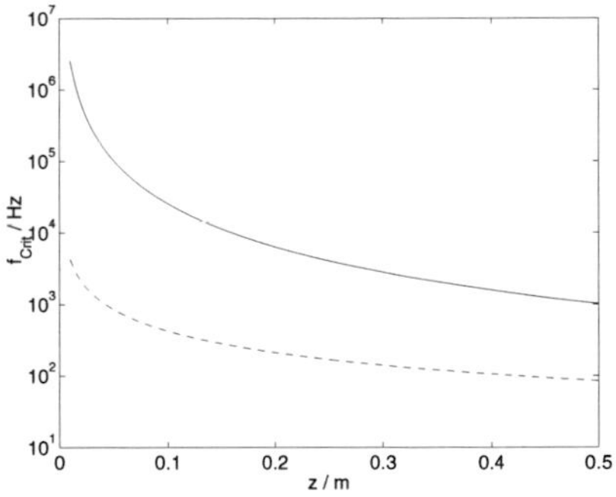

Figure 2.10.: Critical frequency of the Magnitude Model at $\theta = 90°$ (- -) and Complex Model at $\theta = 0°$ (−), $d = 3$ m.

means that calculation results for the coherent combination of either one with another source will not be valid above that frequency.

One may argue that the measurement should rather be performed around the center of the HF unit, choosing the HF source point as the reference point. Of course this will eliminate the error for the HF part, but it also doubles the error for the LF part. In that case the spacing is $z = 0.2$ m and the upper frequency limit is $f_{Crit} \approx 213$ Hz. This limit should not be exceeded by the crossover frequency between LF and HF, if the loudspeaker is combined with another sound source in a computational model.

Even though the overall situation was exaggerated by choosing the direction of maximum error, the phase error of the Magnitude Model clearly represents a significant problem for modeling combinations of sources.

When applying the Complex Model, the upper frequency limit for the same spacing $z = 0.1$ m and measuring distance $d = 3$ m (and choosing $\theta = 0°$ where the error is maximal) is about $f_{Crit} = 25.5$ kHz. Evidently, this approach is not only quantitatively superior but it actually puts the relevant limits far beyond the frequency range of interest. Being free to choose the point of rotation almost arbitrarily resolves a whole number of practical problems, including the determination of the exact source locations (acoustic centers) and the challenge of mounting the device mechanically stable for the directional measurement. This is one of the core results of this work.

Notice, however, that the above considerations are principally limited to the far field of the loudspeaker. It was assumed that the measuring distance d and the receiver distance D are large compared to any characteristic length of the system, such as the offset z. These conditions are normally fulfilled in practice and they are commonly used for acoustic predictions. For the sake of simplicity, it was also assumed that the receive distance is large compared to the measuring distance which is true for most applications.

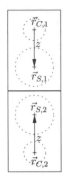

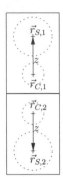

Figure 2.11.: Setup A: source locations spaced 0.2 m apart.

Figure 2.12.: Setup B: source locations spaced 0.6 m apart.

2.1.3. Application to Combined Sources

In the previous section a general method of comparing different data models with regard to their error limits was presented. Using the example of one source and one reference point analytical results have been derived. The thorough analytical treatment of more complex systems, such as interacting loudspeakers with one or more sources each, is complicated and does not provide additional insights. Therefore, in this part numerical results will be presented that illustrate the consequences of omitting phase data. The focus will be on a system that consists of two subsystems each described by a source and a reference point.

Model

Based on the example of the previous section, it is worthwhile to discuss the consequences of choosing the point of rotation at the center of the HF unit like it is commonly practiced. See [97] for a detailed discussion of other cases.

The setup is a stack of two loudspeakers that are arranged in two different ways. In *Setup A* (Fig. 2.11) the LF units are located close to each other. In *Setup B* (Fig. 2.12) the HF units are located close to each other. The transformation from Setup A to Setup B can be understood as a rotation of each subsystem by 180°.

The center of the HF unit is identical with the point of rotation, indicated by $\vec{r}_{C,1}$ or $\vec{r}_{C,2}$. The center of the LF unit is used as the source location, $\vec{r}_{S,1}$ or $\vec{r}_{S,2}$, with an offset from the reference point of $z = 0.2$ m. The two reference points are located at a distance $|\vec{r}_{C,2} - \vec{r}_{C,1}|$ of 0.6 m in Setup A, or 0.2 m in Setup B, from each other. In Setup A the distance between the source points is $|\vec{r}_{S,2} - \vec{r}_{S,1}| = 0.2$ m, and 0.6 m in Setup B .

Now the resulting performance of the combined LF units at frequencies above the critical limit will be discussed. For each setup the results obtained using the Magnitude Model, the Complex Model and the Exact Model are compared. The complex pressure

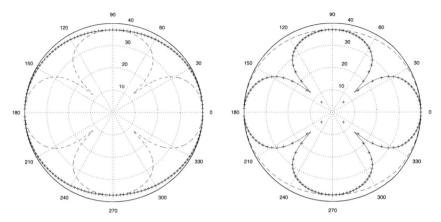

Figure 2.13.: Directional response of setup A at 425 Hz, Magnitude Model (- -), Complex Model (−), Exact Model (+).

Figure 2.14.: Directional response of setup B at 425 Hz, Magnitude Model (- -), Complex Model (−), Exact Model (+).

functions have been computed for the reference point of each subsystem in analogy to equations (2.11), (2.13), and (2.15), thereby using $z = |\vec{r}_{S,1} - \vec{r}_{C,1}|$ or $z = |\vec{r}_{S,2} - \vec{r}_{C,2}|$, respectively. The resulting values were summed up in a complex manner as in eq. (1.4).

Results

To illustrate the results the directional response at a fixed frequency is shown. The measuring distance was chosen to be $d = 3$ m, the receiver is located at $D = 50$ m. The following plots display the modulus of the complex pressure sum for all angles θ. For display purposes the values are presented in dB and normalized to a maximum of 40 dB.

Figures 2.13 and 2.14 show the directional pressure response of Setup A and Setup B, respectively, at 425 Hz. It is obvious that the results obtained from the Exact Model are very close to those obtained from the Complex Model. But one can also see significant deviations when comparing these two with the results from the Magnitude Model.

Comparing the two figures, the Complex Model and the Exact Model exhibit stronger interference effects for setup B, Fig. 2.14. This is expected since the sources are located further apart. On the contrary, the Magnitude Model shows weaker interference effects for setup B. That is because the combined result is computed based on magnitude-only directional data and on reference points being the origins of the wave fronts which are closer in setup B. The directional results of the Complex Model and Exact Model on the one hand and of the Magnitude Model on the other hand basically switch place between setup A and setup B because the source points and reference points switch place.

For the Magnitude Model the critical frequency is $f_{Crit} \approx 213$ Hz. For the Complex Model the critical frequency is $f_{Crit} \approx 6375$ Hz at this measuring distance. It can be seen directly in Figures 2.13 and 2.14 ($f = 425$ Hz) what it means to use the Magnitude Model only one octave above the critical frequency. The pressure values very close to the system axis ($\theta = 0°$), where the error of the Magnitude Model is smallest, are still in acceptable agreement with the Exact Model. However, pressure values off-axis show significant deviations.

It can be recognized that the Magnitude Model cannot accurately represent the chosen configuration of sources and reference points above the critical frequency. Compared with the Exact Model one finds a qualitatively different behavior for the Magnitude Model (see also [97]). As a very practical conclusion one may state that combinations of full-range, two-way loudspeakers whose crossover frequency is significantly above 200 Hz cannot be computed properly using the Magnitude Model.

2.1.4. Measurement Requirements

For the practical application of the above findings one should mention some crucial points of the measurement procedure in order to avoid simple mistakes that can lead to significant errors.

First of all, it is important that the directional measurement data are referenced back consistently to the reference point for any angle θ. It was demonstrated that the true location of the reference point compared to the source does not really matter as long as it is within the desired error range. This is in contrast to what one may believe at first glance, namely that in order to acquire phase data the exact source point must be found and chosen as the reference point. Still, in order to compensate accurately for the propagation delay, both propagation distance and propagation speed must be measured precisely [97], [100].

In the previous section the critical frequency had been introduced as a measure for the upper limit of the valid frequency range that is based on a fixed maximum phase error. For practical applications, simplified expressions for the critical frequency (eqs. (2.25) and (2.26)) can be derived by using the value for angle θ where the error is maximal and fixing the speed of sound c at 340 m/s.

Then, the critical frequency for magnitude-only data is:

$$f_{Crit}^{(Mag)} = 42.5 \frac{1}{z} \text{Hz}, \tag{2.27}$$

where z denotes the offset between source and reference point in m. For any loudspeaker that is intended to be combined with other loudspeakers over the full audio range, the point of rotation needs to be within less than 3 mm relative to the source. For most multi-way systems this means that each transducer should be measured and modeled individually.

Similarly the critical frequency for complex data can be derived:

$$f_{Crit}^{(Com)} = 85 \frac{d}{z^2} \text{Hz}, \tag{2.28}$$

where z denotes the offset between source and reference point in m and d represents the distance used for the acquisition of directional data in m. This establishes a limit of 13 cm for the distance between point of rotation and acoustic source when the measurement radius is 3 m. Modern laboratories use radii of up to 8 m which corresponds to about 20 cm tolerance. This allows choosing the reference point almost anywhere inside the box even for fairly large systems. It also allows measuring multiple transducers together or individually around the same point of rotation. Even more, this result simplifies the very practical problem of choosing the reference point under the constraints of mounting possibilities, mechanical stability or simply work efforts.

Even though in practice combinations of sources can usually be modeled only up to 4 to 8 kHz realistically, the foregoing examples have used 16 kHz as the desired upper frequency limit. That is because results based on frequencies close to the critical frequency are only rough approximations, restricted to qualitative discussion. To be able to utilize the results quantitatively, a significantly lower working frequency than the critical frequency should be used or a smaller phase tolerance than 45°.

2.2. Model of Complex Directivity Point Sources (CDPS Model)

In the previous section it was shown that phase information is crucial when combinations of coherent acoustic sources should be modeled accurately and efficiently. Expanding the data set that describes the directional response characteristics of a loudspeaker or transducer from magnitude-only to complex data requires that the matrix $\hat{A}_{i,k,l}$ in eq. (1.9) becomes complex-valued. The description of an acoustic source in the computational domain by a point source with complex-valued, frequency- and angle-depending radiation amplitude is what will be called here the *Complex Directivity Point Source* or *CDPS* model.

Numerically, the inclusion of phase data requires additional care with respect to the discretization of the spatial and spectral properties of the source. These questions of data resolution and interpolation as well as resulting measurement requirements and related uncertainties of the modeling process will be the concern of this section.

It should also be emphasized that the CDPS model does not necessarily mean that each physically separable acoustic source is represented by a corresponding point source in the model. Rather, a point source in the model may also represent several acoustic sources, such as multiple transducers in a loudspeaker box. In an abstract sense, multiple point sources may also be combined in the model to reproduce the acoustic performance of an acoustic source in reality. The only restriction for these computational images of the real world is that the requirements with respect to data resolution and measurement parameters are met.

Naturally, the model must also satisfy qualitative requirements. If the computational model of the real-world process requires transducers to be controlled individually, e.g. by a DSP chip, these must be represented by individual point sources. Similarly, the transducers should be measured inside the device if effects of the enclosure should be accounted for. Only in this manner the point source approach can include diffraction and baffle effects caused by the box itself (see also Chapter 3).

The CDPS model finds its practical realization in the Generic Loudspeaker Library (GLL) description language which will be presented in Section 2.3.

2.2.1. Processing Phase Data

Processing the directional data of a loudspeaker in order to use them in a computer model requires choosing a sufficient resolution for the sampling step $A(\theta, \phi, \omega) \rightarrow A(\theta_j, \phi_k, \omega_l) =: \hat{A}_{j,k,l}$ and a related, appropriate interpolation function $g(\hat{A}_{j,k,l}, \theta, \phi, \omega) \rightarrow A(\theta, \phi, \omega)$, see eq. (1.9). Now the necessary conditions will be derived in order for the phase interpolation function to be meaningful.

As stated earlier, the measured phase response of a loudspeaker in a certain direction is largely dominated by propagation delays in the form of

$$\psi(\omega) = -\omega t + \psi_0(\omega), \tag{2.29}$$

where t represents the propagation time and ψ_0 is the (small) inherent phase. This is especially true if only a single transducer is measured and it is offset from the POR.

The above form indicates that interpolating phase data separate from magnitude data makes sense physically because the delay time can be considered the underlying physical quantity. Unfortunately, the exact propagation time t in eq. (2.29) is seldom known and mostly not even well defined. Loudspeaker phase data are normally derived from the Fourier transform of a time domain impulse response measurement. These phase data are given in a wrapped format which means that phase values are limited to a circular interval between 0 and 2π. This can lead to ambiguities when interpolating between neighbor data points if the differences between phase values are large. It is then unclear how to resolve discontinuities. The extraction of the propagation time from given wrapped phase data or the unwrapping of phase information from wrapped data in order to obtain a continuous function that can be interpolated is an ambiguous, error-prone process, too [49].

Local Unwrapping

This problem can be largely avoided by requiring that phase differences between directionally or spectrally adjacent data points are small. Formally, this corresponds to imposing a limiting condition on the acquired phase data such as:

$$|\psi_m - \psi_n| < \frac{\pi}{2}, \tag{2.30}$$

where m and n indicate any pair of phase values that are subject to a joint interpolation. It is denoted that the modulus above is meant to be taken on the unit circle where the maximum possible phase difference between any two points is π (Fig. 2.15).

2. Modeling Small Sound Sources

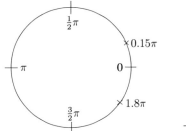

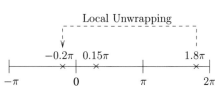

Figure 2.15.: Wrapped phase data points (×) on the unit circle. Condition (2.30) is fulfilled.

Figure 2.16.: Phase data points (×) in the linear domain. Before and after local unwrapping.

Not only for comparing phase values reasonably according to eq. (2.30), but also for interpolating between neighbor phase data points one must ensure that all phase values are located in the vicinity of each other. If eq. (2.30) is fulfilled, phase data points reside close to each other on the unit circle (Fig. 2.15). Any gaps formally greater than $\pi/2$ can be eliminated by *locally unwrapping* the data. This is accomplished by selecting one value and shifting all other values by multiples of 2π, so that all of them are located inside an intervall of width $\pi/2$, see Fig. 2.16. After that interpolation or averaging, may it be spectral or spatial, can be applied. This sequence is repeated for each set of data points to be processed. It should be emphasized that depending on the neighbors considered at a time, a particular data point may assume different absolute phase values.

In practice, the locality of the unwrapping process must be considered a key solution for dealing with measured phase data computationally. In contrast, algorithms based on global unwrapping are prone to errors caused by local noise or other measurement uncertainties at single data points [49].

2.2.2. Data Resolution

Assuming that the delay term in eq. (2.29) is dominant and combining it with eq. (2.30) one can derive a limit for the maximum propagation delay that can be included in the phase response. This in turn leads to conditions with respect to the spatial and spectral resolution of the directional measurement when considering a setup where the acoustic source is offset from the point of rotation [49].

Frequency Resolution

Any phase differential between two frequencies ω_m and ω_n due to the propagation delay t can be characterized by

$$\Delta\psi = (\omega_m - \omega_n)t, \tag{2.31}$$

2. Modeling Small Sound Sources

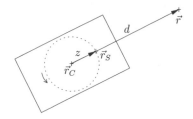

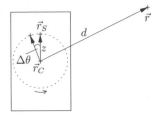

Figure 2.17.: Phase difference between adjacent frequency data points is maximal when the source $\vec{r}_S$ is located on the line through reference point $\vec{r}_C$ and measurement point $\vec{r}$.

Figure 2.18.: Phase difference between adjacent angular data points is maximal when the source $\vec{r}_S$ is close to the perpendicular of the line through reference point $\vec{r}_C$ and measurement point $\vec{r}$.

and using the frequency resolution $\Delta f = (\omega_m - \omega_n)/(2\pi)$:

$$\Delta\psi = 2\pi\Delta f \frac{z}{c}, \tag{2.32}$$

where c is the speed of sound, $z = ct$ the offset of the source. Condition (2.30) is equivalent to a maximum slope of the phase response, which in turn corresponds to a maximum time offset for any measurement direction. This yields a critical offset:

$$z_{crit} \approx \frac{c}{4\Delta f}. \tag{2.33}$$

For example, the bandwidth of a 1/24 octave band around 8 kHz is approximately Δf = 230 Hz. If the condition given by eq. (2.30) was to be fulfilled, the maximum delay time included in the phase data had to be about $t = 1$ ms and $z_{crit} = 0.3$ m. It will occur when the source is located on the line through the point of rotation and the measurement point, on either side of the POR (Fig. 2.17). If data to be interpolated are normalized to the on-axis response the position on the back side determines the maximum delay and the maximum allowed offset z is halved, so that $z_{crit} = 0.15$ m. Of course, larger time differentials can still be covered by higher frequency resolutions.

Angular Resolution

The phase differential between spatially adjacent data points due to different propagation delays t_m and t_n can be characterized by

$$\Delta\psi = \omega(t_m - t_n). \tag{2.34}$$

For a set of points employing an angular resolution of $\Delta\theta$, the maximum time differential $\Delta t = t_m - t_n$ will occur close to the point where the connecting line between the acoustic

2. Modeling Small Sound Sources

source and the point of rotation is perpendicular to the connecting line between the POR and the measuring location (Fig. 2.18). In that case one can approximate for small $\Delta\theta$,

$$\Delta\psi = \frac{\omega}{c} z \sin(\Delta\theta), \qquad (2.35)$$

where z is the offset of the acoustic source from the point of rotation[3]. In order to agree with condition (2.30) the distance of the acoustic source from the POR must not be greater than

$$z_{crit} \approx \frac{c}{4f\sin(\Delta\theta)}. \qquad (2.36)$$

For example, in order to comply with eq. (2.30) at 8 kHz at an angular resolution of $\Delta\theta = 5°$, the acoustic source should be located not further away from the POR than about 0.12 m. Of course, increasing the angular resolution allows for a larger spacing.

2.2.3. Data Management

In addition to the resolution requirements given above, experience has shown that a number of practical measures and management techniques can further improve overall data consistency and modeling accuracy. The most important ones are listed here briefly.

On-Axis Data and Directional Data

In practice, data management is simplified and interpolation results are improved by storing the on-axis response of the source separate from the directional data that are normalized to it. Normalizing directional data removes absolute propagation delays, eq. (2.29), from the phase data and reduces the complexity of the magnitude data. Both effects usually allow for easier interpolation later on.

Another practical advantage of this method is that small changes to the loudspeaker, such as to its equalization settings, can be incorporated into the overall data set by simply re-measuring and replacing the data of the on-axis direction only rather than that of the entire set. Of course, this is only applicable if the three-dimensional radiation pattern is not affected by the change.

On-Axis Re-Normalization

When acquiring directional data, measurements are taken for a number of directions. These are normally given by a grid of points based on the meridians and parallels of a spherical surface around the loudspeaker. Most often, one of the poles of the grid is chosen to be in the on-axis direction of the loudspeaker. This is useful to increase the resolution in those directions where prediction accuracy is most important.

A three-dimensional directional measurement involves rotating the loudspeaker around two axes. The most common method consists of first turning the loudspeaker step by step from the front to the back side. Then the loudspeaker is returned to its original

---

[3]For this approximation it was also assumed that z is much smaller than the actual measuring distance.

position, tilted by a certain angle around its axis and again rotated front-to-back. This yields a number of redundant measurements for the on-axis direction as well as for the back side.

Since most directional measurement series take several hours, the measured directional response of the loudspeaker may drift slightly due to changing ambient parameters, even in a controlled environment. Also, the mechanical rotation of the loudspeaker around its axis may lead to small deviations of its perceived on-axis radiation relative to the fixed microphone position. For these reasons it has proven useful to normalize each front-to-back measurement data set to the corresponding on-axis measurement rather than to the very first. Slow drifts over the course of hours can thus be avoided. Drifting parameters during a single front-to-back run are less critical since each run takes only some minutes.

Adaptive Data Resolution

Depending on the type of loudspeaker and its radiation characteristics, lower data resolutions may be sufficient for some parts of the grid compared to others. Often the back side of the loudspeaker is less important so that a rougher spectral or spatial resolution is acceptable. For some loudspeaker types the angular resolution of the meridians can be chosen differently from that of the parallels due to certain symmetries.

Naturally, one can always measure all data at the highest resolution required. However, this may create substantially longer measurement times as well as significantly larger data sets. These require more processing memory and performance in turn, without necessarily much gain in accuracy.

Delay-Reduced Phase Data

Even though it is often difficult to extract the exact propagation delay from processed phase data, it is advantageous to reduce the phase response similar to eq. (2.29) by a broadband, effective delay time that is stored separately.

This kind of treatment has primarily memory reasons. After converting from the possibly high frequency resolution of the raw measurement data to a lower frequency resolution used for the modeling data, the phase response should still comply with eq. (2.30). First unwrapping phase data and then reducing the frequency resolution is possible but requires 4 to 8 bytes per data point. Calculating an average delay time and subtracting it from the phase response before reducing the frequency resolution is a good way to keep the slope of the phase data small. That allows storing wrapped phase data in rad for which it is quite efficient to use 2-byte integer values for each data point. In combination with a single delay value and 2 bytes for each magnitude data point this structure facilitates high-resolution data sets easily.

2.2.4. Errors

Unfortunately, commercially available acoustic modeling software packages do not provide much information on the uncertainty of the simulation results. Although it is difficult to estimate errors related to model data input by the user, one should be able to derive prediction uncertainties arising from errors in the loudspeaker measurement data, for example. In this section, only an overview is given over the sources of error that are most significant in practice.

First of all, the consistency of the measurement data with respect to on-axis and directional response is important. Redundant measurements during the acquisition procedure such as mentioned above or explicit tests can be used to estimate the reproducibility of the measurement data. Systematic errors may occur as well, most commonly these are related to the determination of the point of reference relative to the microphone or to slight mechanical instabilities of the mounting of the loudspeaker.

Another important issue is the variation of performance among several loudspeaker samples of the same type. The spread of the production can lead to severe systematic errors if the representative unit chosen for the measurement of directional data is far away from the mean of set. For this reason, it is strongly recommended to either take loudspeaker measurements of the so-called *golden unit*, that is, of the sample that is closest to the mean of set. Alternatively, several samples should be measured in order to estimate the spread directly. Of course this still poses questions about the modeling results in relationship to the loudspeaker samples installed in reality later. These can be off from the mean of set, as well. The effects of sample variation on line array performance will be discussed in Chapter 3.

Data processing errors may occur as well, including the undersampling of the loudspeaker directional performance, the lossy conversion of raw measurement data into the storage format of the computer model and the interpolation of stored data in order to reproduce the radiation functions that are continuous in space and frequency. Care must also be taken when using large unwrapped phase values. Complex summation only considers the wrapped part of the phase. Therefore using floating point precision numbers with comparatively few valid digits can cause numerical errors.

Conceptually, it must also be clear that the point source model cannot reproduce wave-based effects, such as diffraction. The acoustic support of a neighbor loudspeaker cabinet especially at low frequencies is normally neglected in this model, as well. Some of these issues will be discussed in the next chapter when looking at large sources.

These and potentially other errors will propagate through the calculations and will be reflected in the end result, e.g. with respect to predicted overall sound pressure level or propagation time. Until now only rules of thumb exist for estimating uncertainties of the final simulation results. It will be the goal of the next generation of simulation software to provide the end user with better error estimates.

2.3. The GLL Description Language and Data Format

The Generic Loudspeaker Library (GLL) is a file format and description language for the mechanical, electronic and acoustic properties of loudspeakers. It can be considered the formal and practical realization of the CDPS model[4]. A thorough introduction into the GLL data format is given in [48]. The detailed discussion of the technical aspects of the GLL format is beyond the scope of this work. An abstract overview is provided here before this chapter concludes with a number of comparisons between measurement and simulation results. All of these were derived using GLL data for the loudspeakers and systems considered.

2.3.1. Overview

The GLL format is generally designed to describe loudspeaker systems and acoustic sources of any kind. This expressedly includes

- conventional line arrays,

- column loudspeakers and steered columns,

- loudspeaker clusters and arrays,

- multi-way loudspeakers.

All of these are difficult to represent by a simple data table that consists of a single data set of directional response data (see also Section 1.3). From this perspective, there are several distinct features of the GLL format. They establish a significant step forward in the modeling of loudspeakers compared to conventional, tabular data formats:

- The GLL facilitates complex-valued directional transfer function data in high resolution according to the requirements of the CDPS model.

- The loudspeaker or loudspeaker system represented by the GLL can consist of one or multiple acoustics sources, each with their own set of data.

- The GLL is an object-oriented description language. In the GLL format there are equivalent objects for each component of a real-world loudspeaker system.

- The user can configure the loudspeaker or loudspeaker system in the GLL model similar to the real world, e.g. regarding crossover settings and mounting choices.

These advancements provide flexibility and scalability as well as convenience of use. They also reduce the compromises required for the recreation of the loudspeaker model in the software domain. This is aided by the fact that constraints historically originating from limited memory and computing power are of less concern today [48].

---

[4]The current version of the recently published AES standard AES56 [51] also accounts for the requirements of the CDPS approach and suggests using directional transfer function data in high-resolution.

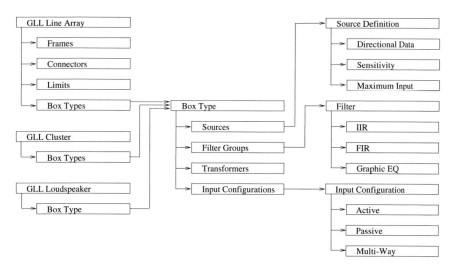

Figure 2.19.: Overview of GLL object hierarchy.

2.3.2. General Structure

In the GLL model, there are three fundamental, hierarchical levels of data.

1. On the lowest level, there is the *source definition*. It determines an acoustic source or electro-acoustic transducer by its sensitivity, its directional data, its on-axis frequency response, its maximum input voltage, as well as its other properties. The data of an acoustic source is stored in a GSS (Generic Sound Source) file that can be referenced and included by one or multiple GLL files.

2. A *box type* represents the loudspeaker cabinet with electronic inputs, acoustic outputs, and a processing unit that connects them. A box type can be composed of one or several sound sources (acoustic outputs), each of which refers to a source definition. A box type also has one or more input configurations that define how many external inputs are available. The input matrix of each configuration then defines how the electronic inputs are linked to the acoustic outputs. Each node of the input-output matrix can be populated with filter settings, such as for crossover, mixing or equalization. This scheme allows implementation of the most simple passive loudspeaker as well as of active or passive multi-way loudspeakers with switchable inputs, and even digitally steered columns.

3. One or several box types can be combined into a group of boxes, such as a *line array* or a *cluster*. A cluster in the sense of the GLL is considered as a simple set of boxes each with its own location and orientation as well as an assigned box

43

type. For line arrays, additional mechanical information regarding the available splay angles between cabinets as well as available rigging and groundstack frames must be defined. Mechanical limits, such as regarding the maximum number of cabinets or the maximum weight, can be included as well.

An overview of the object hierarchy in the GLL concept is shown in Fig. 2.19. It should be noted that due to the object-oriented composition new types of objects can be added and existing ones can be modified without revising the entire data format.

2.3.3. Practical Implementation

In practice the creation of the GLL model for a loudspeaker or loudspeaker system involves several steps. Acoustic measurement data as well as mechanical and electronic data must be collected as outlined before. These data are then combined in a so-called GLL project which is subsequently compiled into a single GLL file (Fig. 2.20). Additionally, this file can contain different system and filter presets as well as informational data, such as a user manual and technical drawings. The GLL file can also be "authorized" by the creator of the GLL in order to indicate that it is original and approved data. The final GLL data file is then distributed to software users.

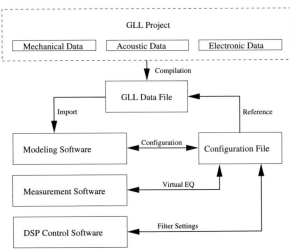

Figure 2.20.: Exchange of GLL modeling and configuration data.

Design Phase and Use Phase

It should be emphasized that based on the above concept there are two phases, the creation phase and the use phase, similar to writing software code at design time and executing the code at run time. First the GLL is assembled and compiled by the designer. In this step all of the information is included about what the end user will see and what changes he is permitted to make to the loudspeaker system. In the second phase the end user loads the GLL into a modeling software where he is then able to configure the loudspeaker model according to the degrees of freedom implemented earlier by the GLL creator. This philosophy is very similar to the actual usage of the real-world device.

In the room model, the end user will then configure the GLL model according to the needs of the project. Information about this GLL configuration can be saved and loaded in order to exchange data between different loudspeakers in the project, between different projects, as well as between different modeling platforms. Configuration files can be exchanged even with measurement software and DSP control software if supported. Fig. 2.20 illustrates this process.

DLL Extensions

Sometimes it may be necessary to expand the GLL functionality in a specific way, such as with respect to a product-specific user interface for the configuration of the loudspeaker model or with respect to proprietary filtering algorithms for beam-steering. For this purpose the GLL can be complemented by a separate software module, e.g. a DLL (Dynamic Link Library), which allows programming the necessary extensions.

Geometrical Visibility Tests

For GLL models with many transducers, the performance of shadowing and raytracing calculation may be low due to the large amount of sources. Practical evaluations have shown (see e.g. [101]) that grouping several sources for geometrical visibility tests relative to a common, so-called *virtual center point* allows increasing computation speed without measurable loss in prediction accuracy.

Figures 2.21a and 2.21b illustrate this concept. In a first step, acoustic sources are grouped. For each group a virtual center point is defined. The geometrical visibility test is then performed for each virtual center point (Fig. 2.21a). Finally, all acoustic sources that belong to a visible center point are summed at the receive location (Fig. 2.21b).

Compared to the typical resolution of acoustic room models as well as with respect to first-order diffraction effects a grouping diameter of about 0.5 m has turned out to be sensible. Grouping levels and virtual center points can be determined automatically by the modeling software or adjusted by the user as needed.

Description Language

The GLL project (see also Fig. 2.20) consists normally of raw measurement data in a binary or text format as well as of the project file in a text format. The format of the project file follows a simple key-value based, hierarchical scheme, similar to XML. It can be edited manually or using the EASE SpeakerLab software [102]. The EASE SpeakerLab software is also used to generate the compiled GLL file. Based on the open text format, other modeling platforms may also provide editing functions and

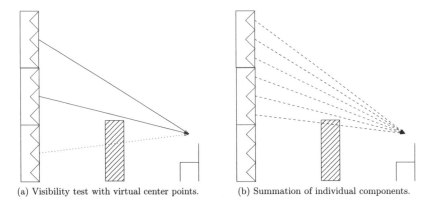

(a) Visibility test with virtual center points. (b) Summation of individual components.

Figure 2.21.: Geometrical visibility tests. Triangles indicate individual sound sources; sources are arranged in groups indicated by rectangles.

compilation into their target file format for data distribution. Fig. A.1 in Appendix A.2 shows an example of a GLL project text file. A detailed description of the format as well as more extensive examples are given in [48].

2.4. Validation of Loudspeaker Models

In the following part, in order to demonstrate the accuracy and practical usefulness of the physical CDPS model in combination with the numerical GLL description language, a number of different examples will be discussed that compare measurement results of radiated sound fields with modeling results. With respect to previously published data as in [49], [100] the selection shown here is focused on the contents of this thesis. Also, the data compiled here were obtained from new simulation runs at higher resolution and with refined algorithms. Consequently, there will be minor deviations compared to previous publications.

All of the following measurements were acquired in high resolution by means of an FFT-based measurement system and are based on impulse response or complex frequency response data with linear spacing. All directional measurements were performed in 3D, at 5° angular resolution and on a half-sphere or full-sphere grid around the loudspeaker. The measurement distance used was about 5 to 8 m which puts the receiver approximately in the far field of the loudspeaker unless stated differently. Measurements of individual transducers were always made with the transducer in the loudspeaker box.

Simulation runs were performed with EASE SpeakerLab [102] and corresponding GLL data files for each setup. For better comparison most modeling data were computed at 2.5° angular resolution.

Results are largely presented as directional response plots for the vertical plane along the loudspeaker axis or for the horizontal domain looking from above the loudspeaker. These polar plots are normalized to the maximum value and shown for a range of 40 dB. The on-axis direction (0°) is to the right of the plot; the direction of 90°, upward in the plots, corresponds to the upward direction for the vertical plane and to the left of the loudspeaker for the horizontal plane.

In the following comparisons particular emphasis will be put on:

- Accuracy of the magnitude model and the complex model.

- Usage of different points of rotation and their influence on the prediction accuracy.

- Different loudspeaker setups and configurations.

- Potential causes for systematic deviations between measurement and simulation.

2.4.1. Two-Way Loudspeaker

As a first example, the performance of a simple two-way loudspeaker, the Renkus-Heinz PNX121T, will be investigated. The on-axis frequency response is displayed in Fig. 2.22; it indicates a crossover frequency of about 1.6 kHz. Fig. 2.24 shows the vertical directional response of the horn and of the woofer at that frequency, respectively. The system was measured as a whole and then compared with the software reproduction based on different measurements of the horn (HF) and the woofer (LF), Fig. 2.23.

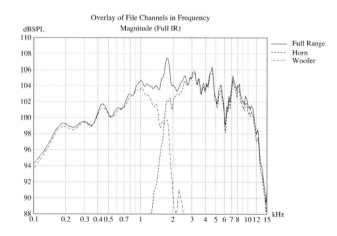

Figure 2.22.: Frequency response of PNX121T, HF unit (- -), LF unit (-.-) and combined (−).

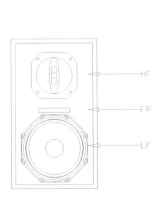

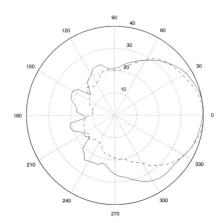

Figure 2.23.: Two-way loudspeaker PNX121T, with indications for horn (HF), port (FR), and woofer (LF).

Figure 2.24.: Directional response, vertical plane, of horn (−) and woofer (- -) for 1600 Hz at 1/3 octave bandwidth.

Aside from the horn and the woofer also the port[5] (FR) served as the point of rotation. The center of the horn is located approximately 0.15 m above the port, the center of the woofer approximately 0.17 m below the port.

Fig. 2.25 shows a comparison of measurement, prediction using complex data, and prediction using magnitude-only data at the crossover frequency. When using the individual centers of horn and woofer for the point of rotation, respectively, there is little difference among all three data sets (Fig. 2.25a). For the front half, the match is very good whereas for the back half the magnitude-only data set shows little deviations.

In contrast, when employing measurement data that were acquired by rotating the woofer or the horn about the port (2.25b), the error of the magnitude model is expectedly large, because it cannot account for the spatial offset between acoustic source and point of rotation. As explained in Section 2.1, complex-valued directional data include that information directly in the phase data. Therefore the match with the measurement is equally good compared to the first case. This loudspeaker will be the basis for the considerations in the following Section 2.4.2.

In order to demonstrate the accuracy and flexibility of the complex model, the loudspeaker was slightly modified by applying additional filtering to its inputs. In one setup, the horn was attenuated by 12 dB in both simulation and practice. The resulting vertical response at the new crossover frequency of 2 kHz is illustrated in Fig. 2.26a. In another setup, the woofer was delayed by about 0.334 ms which equates to steering the main lobe downward by approximately 20°, in the crossover range. Also in this case a

---

[5]This choice is not essential for the following analysis. Often the geometrical center is also chosen for measuring loudspeaker cabinets.

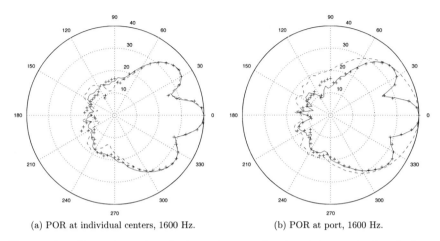

(a) POR at individual centers, 1600 Hz.　　　(b) POR at port, 1600 Hz.

Figure 2.25.: Comparison of predictions using complex data $(-)$, magnitude-only data $(- -)$, and measurement $(+)$, at 1/3 octave bandwidth.

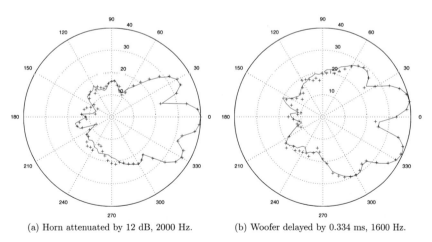

(a) Horn attenuated by 12 dB, 2000 Hz.　　　(b) Woofer delayed by 0.334 ms, 1600 Hz.

Figure 2.26.: Comparison of predictions using complex data $(-)$, and measurement $(+)$, at 1/3 octave bandwidth.

good match between model and reality was found, Fig. 2.26b. Having introduced the concept here, the practical application of utilizing the GLL model for optimization of crossovers will be discussed in Section 2.4.3.

2.4.2. Loudspeaker Clusters

After the introduction of the single loudspeaker box in Section 2.4.1, the interaction of two such boxes in different configurations will now be investigated. Measurements and simulations of a stack of two two-way loudspeakers arranged horn-to-horn (Fig. 2.27a) and woofer-to-woofer (Fig. 2.27b) as well as of a side-by-side setup (Fig. 2.27c) will be compared. The results demonstrate that the complex model (CDPS) is superior over the magnitude model and that the computation accuracy depends on configuration, frequency range, and data resolution.

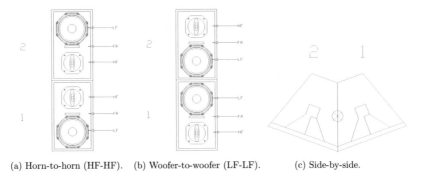

(a) Horn-to-horn (HF-HF).   (b) Woofer-to-woofer (LF-LF).   (c) Side-by-side.

Figure 2.27.: Modeled and measured setups of two two-way loudspeakers.

HF-HF Setup

In the HF-HF setup the two loudspeakers are stacked horn-to-horn (Fig 2.27a). Fig. 2.28 shows the calculation results for the complex model and for the magnitude model as vertical polar plots at different frequencies, namely below, around, and above the crossover frequency of 1.6 kHz. Below and above crossover, only the LF or the HF transducers, respectively, are active. In the crossover range all four sources interact with each other. Fig. 2.29 displays the absolute difference between model results and measurement data, averaged over an opening angle of ±60° and as a function of frequency. In this study three different simulation methods are distinguished:

- Calculation based on a single full-range data set for each box, with the point of rotation at the port.

- Calculation based on a single full-range data set for each box, with the point of rotation at the horn.

- Calculation based on a separate data set for the woofer and for the horn, each with the point of rotation at the respective center.

As one can see, the complex model yields very accurate results with only small deviations. Differences are mostly within a range of ±1 dB for the front side. They are slightly larger for the high-frequency range. This is likely due to the fact that at high frequencies the superposition model has an increased sensitivity against small positioning errors and greater source directionality.

Compared to the other two data sets the full-range data set that was measured about the port shows a slightly smaller mean deviation for frequencies below the crossover frequency (Fig. 2.29a). A look at the corresponding directional plots indicates that this offset is likely insignificant and may be due to small differences in the mounting or the environmental parameters during the measurement. On the other hand, above crossover the same model shows an increased mean deviation that is approximately 1-2 dB greater than for the other data sets. This difference is also visible in the polar plot. Part of it is probably caused by the 15 cm offset of the POR from the horn. At 4 kHz the corresponding phase error is of the order of 10°. The resulting small angular shifting of minima and maxima in the directional patterns, as in Fig. 2.28e, will lead to relatively large level differences in Fig. 2.29a because the compared curves have steep slopes as a function of angle.

With respect to the magnitude model, among all three setups the version based on individual sources has by far the smallest deviations from the measurement. The largest error of this most accurate magnitude-only data set is in the crossover frequency range. This could be explained by the fact that in this range the phase relationships between low-frequency and high-frequency units play a role. But by definition the magnitude model neglects the inherent phase response and thus differences between the transducers. In contrast, below and above the crossover frequency only the phase difference between the LF units or the HF units, respectively, matters. In this frequency range the phase difference at the receiver is generally small since the directional data for the involved sources are the same and investigations take place in the far field where the receive angle is roughly the same for both transducers. Neglecting phase data under these conditions will cause only a small error in the result.

The magnitude model based on full-range data with the point of rotation at the horn provides expectedly good results above the crossover frequency, where the HF units dominate the vertical response. Interestingly, the version measured about the port yields fairly good results in the crossover range. This is supposedly so because in this frequency range the location of the port seems to establish an average acoustic center between the horn and the woofer.

Overall the magnitude model shows significantly greater deviations than the complex model whenever the POR was not located at the center of the active source during the measurement. It also shows larger errors when different types of sources are combined,

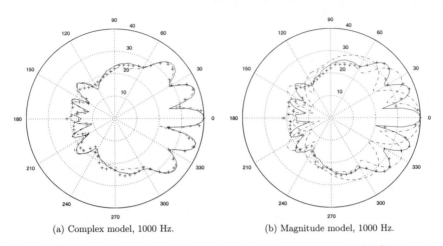

(a) Complex model, 1000 Hz.　　　　(b) Magnitude model, 1000 Hz.

Figure 2.28.: Comparison of predictions using two full-range sources and POR at port (-.-), two full-range sources and POR at horn (- -), four sources (−) and measurement (+), HF-HF setup, at 1/3 octave bandwidth.

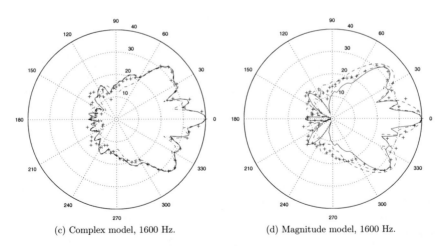

(c) Complex model, 1600 Hz.　　　　(d) Magnitude model, 1600 Hz.

Figure 2.28.: *Continued.*

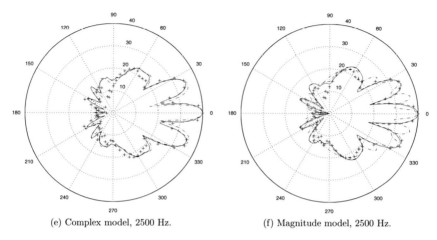

(e) Complex model, 2500 Hz.   (f) Magnitude model, 2500 Hz.

Figure 2.28.: *Continued.*

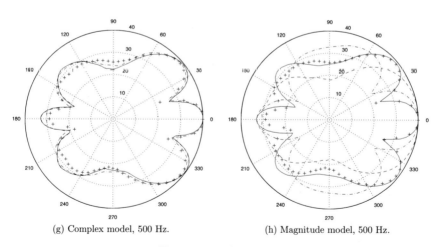

(g) Complex model, 500 Hz.   (h) Magnitude model, 500 Hz.

Figure 2.28.: *Continued.*

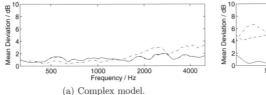 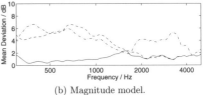

(a) Complex model.   (b) Magnitude model.

Figure 2.29.: Mean deviation of prediction results using two full-range sources and POR at port (-.-), two full-range sources and POR at horn (- -), and four sources (−) relative to measurement, HF-HF setup, averaged over ±60°.

such as LF and HF unit. For the remainder of the parameter space the accuracy of the magnitude model is comparable to the complex model.

In this section, and also in the following comparisons, one point is remarkable regarding the correlation between measured and predicted directivity. In many cases the measured data seem to be slightly different with respect to the shape and extent of the maxima and minima; especially the maxima are broadened and the minima seem not as deep as in the prediction. This can be explained by the fact that in the real world there is always a small amount of random averaging involved that is, for example, due to moving air, diffraction effects, and the finite size of the microphone. On the contrary, the computation is usually artificially precise and may exhibit poles that cannot be measured by any normal technical means. Therefore extreme values appear slightly smoothed in measurement compared to prediction.

Related to that there is another potential source of error, namely the evaluation of the vertical directional response based on normalizing to the maximum. If the measurement does not capture the maximum level as accurately as the modeling output due to above effects or slight angular positioning errors, the level for other directions relative to the maximum is increased erroneously.

One can also notice another typical deviation that occurs here as well as for the other setups with multiple loudspeakers. Especially in off-axis direction around 60° to 120° there are systematic differences in the level and the location of the lobe structure. They appear primarily in the low- to mid-frequency range. These deviations seem to be caused by wave-based coupling and diffraction effects that originate from the adjacent loudspeaker box. Since the GLL models in this chapter are based on point sources that were measured for a single box, such effects cannot be accounted for. This issue will be discussed in greater detail in Chapter 3 when the focus is on line array modeling.

With respect to the back side of the loudspeaker or loudspeaker cluster a pattern in the deviations can be recognized, as well. Not unexpectedly, neither the complex model nor the magnitude model can accurately reproduce the rearward radiation behavior. On the one hand, the loudspeaker is usually mounted at its back side. Naturally, these physical connections will change the directional response. Also, when the back side of the loudspeaker faces the microphone, normally its front side faces the measurement robot or turning construction. Depending on the front-to-back ratio of the loudspeaker's radiation level, sound reflected by the robot may enter into the measurement as a noise floor. Naturally, aforementioned diffraction and shadowing effects by the adjacent loudspeaker cabinets will also affect the back-side measurement.

Another interesting effect concerning the back side can be recognized: Mostly, the magnitude model shows a more pronounced lobe structure than the complex model. This can be explained by the fact that by considering only phase contributions based on the propagation time, the rearward sound is more coherent in the magnitude model than it is in reality. Including measured phase data with the complex model seems to establish a level of randomness that leads to more realistic results when sources are combined.

The last Figure 2.28h shows the results for the magnitude model at 500 Hz. For the given loudspeaker the critical frequency for the woofer is about 150 Hz when the POR is at the location of the horn and about 300 Hz when the POR is at the port. This is reflected by the vertical directional response close to critical frequency. It shows a better qualitative agreement between measurement and magnitude model when using the port as the point of rotation.

LF-LF Setup

The LF-LF setup consists of two loudspeakers that are stacked woofer-to-woofer (Fig. 2.27b). Calculation results are shown in Fig. 2.30. Again, the vertical directional response of the complex model and of the magnitude model is displayed. The same frequencies and parameters were used as for the HF-HF setup. Fig. 2.31 displays the average deviation of measurement and model over an opening angle of $\pm 60°$.

The findings are very similar to the horn-to-horn setup, as well. The complex model yields good results for the different versions of data although the setup with two full-range sources measured about the port shows the largest deviations in comparison. These are mostly within a range of ± 2 dB for the front side.

In contrast to the complex model, the magnitude model can be considered as satisfyingly accurate only if individual sources are used which are measured about their respective centers. Nonetheless, all of the complex model versions seem to be at least as accurate or better.

It is worth pointing out that the limits of the measurement resolution for the combined loudspeakers appear to be reached at about 2500 Hz where the lobe structure is only barely resolved. To further emphasize that, the complex model is also compared at the frequency of 4000 Hz in Fig. 2.32. Here, the maxima of the directional response function of the cluster are spaced approximately 2.5° apart. At this frequency, the 5° measurement data show highly variable results due to sampling errors. The simulation results sampled exactly in 5° intervals even show an entirely wrong curve for the front section due to systematic undersampling. This demonstrates that, for directional measurements of a simple loudspeaker arrangement like this, an angular resolution of 5° is by far not enough if accurate prediction results are desired for frequencies above 2 kHz.

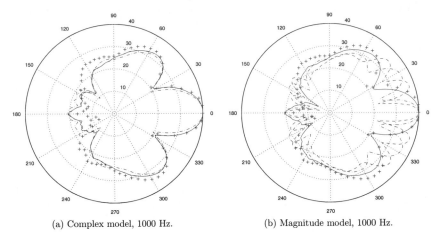

(a) Complex model, 1000 Hz.   (b) Magnitude model, 1000 Hz.

Figure 2.30.: Comparison of predictions using two full-range sources and POR at port (-.-), two full-range sources and POR at horn (- -), four sources (−), and measurement (+), LF-LF setup, at 1/3 octave bandwidth.

While a high angular measuring resolution may solve the problem described here, this is not recommended. For applications in simulations it is obvious that rather than measuring the cluster system at an angular resolution much higher than 5°, one should instead consider measuring the individual sources involved at 5° and using them in combination.

It should be remarked that in Figs. 2.29 and 2.31 only the relevant frequency range is shown. For the above reasons, at higher frequencies the angular resolution of 5° for the measurement data points is not high enough to allow for a useful comparison with the prediction results. At frequencies below ca. 350 Hz off-axis diffraction effects start dominating the average deviation of the model from the measurement, and the computed difference depends strongly on the exact limits of the averaging interval.

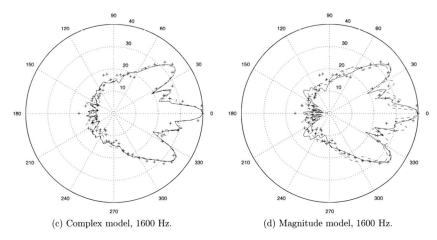

(c) Complex model, 1600 Hz. (d) Magnitude model, 1600 Hz.

Figure 2.30.: *Continued.*

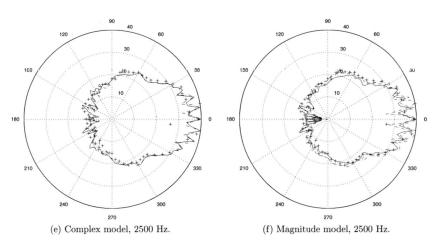

(e) Complex model, 2500 Hz. (f) Magnitude model, 2500 Hz.

Figure 2.30.: *Continued.*

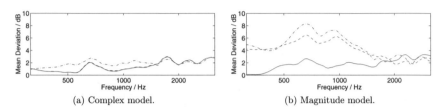

(a) Complex model. (b) Magnitude model.

Figure 2.31.: Mean deviation of prediction results using two full-range sources and POR at port (-.-), two full-range sources and POR at horn (- -), and four sources (−) relative to measurement, LF-LF setup, averaged over ±60°.

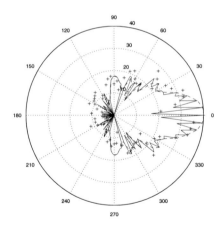

Figure 2.32.: Comparison of predictions at angular resolution of 2.5° (−), resolution of 5° (- -) and measurement at 5° (+), LF-LF setup, complex model using 5° CDPS data, for 4000 Hz at 1/24th octave bandwidth.

Side-By-Side Setup

The side-by-side setup consists of two loudspeakers arranged horizontally at an opening angle of 40°, see Fig. 2.27c. Interaction among sources now takes places in both the horizontal and the vertical domain. Fig. 2.33 displays the horizontal directional response around crossover as well as the vertical directional response at the crossover frequency of 1600 Hz, for the complex model. The results agree with the previous findings. There are no significant differences in the horizontal domain and only small deviations in the vertical domain. These are similar to the ones discussed before.

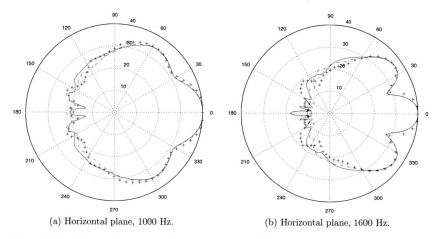

(a) Horizontal plane, 1000 Hz.　　　(b) Horizontal plane, 1600 Hz.

Figure 2.33.: Comparison of predictions using two full-range sources and POR at port (-.-), four sources (−) and measurement (+), side-by-side setup, complex model, 1/3 octave bandwidth.

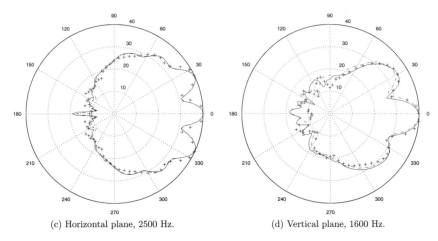

(c) Horizontal plane, 2500 Hz.   (d) Vertical plane, 1600 Hz.

Figure 2.33.: *Continued.*

2.4.3. Crossover Design with Multi-Way Loudspeakers

So far the focus was on combining multiple sources and comparing the calculation result with measurement data for the same setup. The problem becomes more complex when considering systems where additionally filter functions are applied to each source. In this section it will be demonstrated that loudspeakers with electronic filters can be accurately reproduced by the CDPS model [100]. Based on eq. (2.1) the coherent combination of point sources j is given by the complex sum of the corresponding pressure values at the receive location $\vec{r}$:

$$p_{Sum}(\vec{r}, \omega) = \sum_j p_j(\vec{r}, \omega). \tag{2.37}$$

For a linear system and filter transfer functions that do not depend on the rest of the loudspeaker one can simply write

$$p_{Sum}(\vec{r}, \omega) = \sum_j h_j(\omega) p_j(\vec{r}, \omega), \tag{2.38}$$

where $h_j(\omega)$ represents the complex transfer function of the filter that is applied to the acoustic source j. This concept requires that filters are given independently from the data of the acoustic source, either as a modeled or as a measured complex-valued frequency response. If this condition is fulfilled, eq. (2.38) indicates that one may also use the CDPS model to optimize the filter functions h_j based on given p_j in order to achieve a desired result p_{Sum}. Accordingly, not only the applicability of the CDPS model to active and passive loudspeakers will be demonstrated but it will also be shown how the filter settings for a multi-way loudspeaker system can be optimized in the software domain only.

On-Axis Response

As an introductory example the on-axis response of a modeled two-way loudspeaker is investigated. The loudspeaker under test is a Peavey SP-1G with a horn and a woofer. Their measured on-axis transfer functions are shown in Fig. 2.34a. The crossover frequency is located at about 1.5 kHz which is also indicated by the measured filter transfer functions in Fig. 2.34b. The result according to eq. (2.38) is presented in Fig. 2.34c. Clearly, there is a good match between the measured and the modeled on-axis frequency response for the combined system of transducers and filters.

Active and Passive Loudspeakers

The next comparison uses the active and the passive version of the Renkus-Heinz SG/SGX 151 loudspeaker model. The passive setup utilizes an analog filter network that is built into the loudspeaker box. Its crossover frequency is located at about 2 kHz. The active setup was realized using an external DSP controller with the crossover frequency approximately at 1600 Hz. For both setups the transfer functions of the filters were measured. Their magnitude is shown in Figures 2.35a and 2.36a.

After that, full-range directional measurements were made and compared with the prediction that utilizes the individual sources and the crossover filters. The results for the vertical directional response at crossover frequency are shown in Figures 2.35c and 2.36c. Evidently, the agreement is very good. The simulation of the full-range loudspeaker is capable of imaging the real full-range system for different filter settings, both using analog filter networks and DSP-implemented filters. In addition to that, Figures 2.35b and 2.36b show the directivity index (DI) as derived from the full-sphere directional data for both measurement and calculation results. This underlines that the introduced approach can be utilized to predict characteristic figures of loudspeakers as well.

It should be noted that the small deviations towards the lower frequencies may be caused by the increased influence of the time windowing that was slightly different for the full-range and for the woofer measurement. It should also be mentioned that both filter settings are of preliminary nature as the manufacturer indicated.

Compared to active or DSP-controlled loudspeakers, more care must be taken when measuring or modeling the filter networks of passive loudspeakers due to their interaction with other circuit sections and the transducers [100].

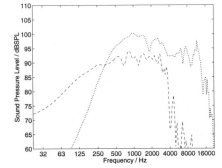

(a) Woofer (- -) and horn (···) measured without filtering.

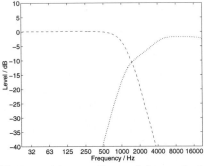

(b) Measured low-pass filter applied to woofer (- -) and measured high-pass filter applied to horn (···).

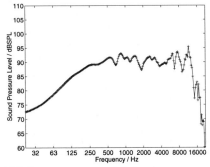

(c) Full-range measurement (+) and response reconstructed from individual measurements (−).

Figure 2.34.: Comparison of measurement with prediction using two sources with filters, on-axis magnitude response at 1/24th octave bandwidth, Peavey SP-1G loudspeaker.

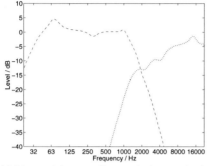

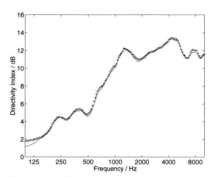

(a) Measured low-pass filter applied to woofer (- -) and measured high-pass filter applied to horn ($\cdots$).

(b) Prediction ($-$) and measurement ($+$) of directivity index DI.

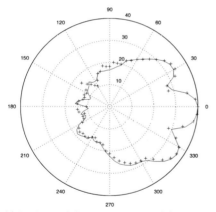

(c) Prediction ($-$) and measurement ($+$) of vertical directional response, 1600 Hz at 1/3rd octave bandwidth.

Figure 2.35.: Comparison of measurement with prediction using two sources with filters, two-way active loudspeaker, Renkus-Heinz SG151.

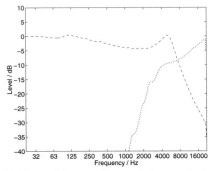

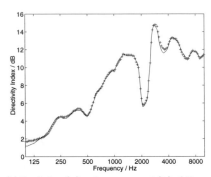

(a) Measured low-pass filter applied to woofer (--) and measured high-pass filter applied to horn (···).

(b) Prediction (−) and measurement (+) of directivity index DI.

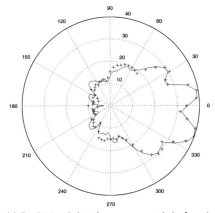

(c) Prediction (−) and measurement (+) of vertical directional response, 2000 Hz at 1/3rd octave bandwidth.

Figure 2.36.: Comparison of measurement with prediction using two sources with filters, two-way passive loudspeaker, Renkus-Heinz SGX151.

2. *Modeling Small Sound Sources*

Crossover Optimization

The process of optimizing the directivity of a loudspeaker system using the CDPS model is not all that different from conventional methods. One of the prime differentiating factors is that once the directional data sets for the individual source (transducer) have been measured the full-range system directivity is calculated. Any arbitrary filter, gain, or delay may be applied to each source individually and the system directivity recalculated. Therefore, the time-consuming process of measuring and re-measuring the system directivity during the design phase can be eliminated.

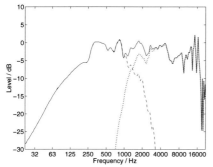 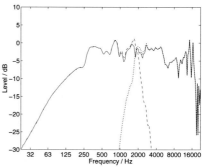

(a) Measured, configuration with initial Linkwitz-Riley filters.

(b) Predicted, configuration with optimized asymmetrical Butterworth filters.

Figure 2.37.: On-axis magnitude response of active two-way loudspeaker for woofer (- -), horn (· · ·), and combined (−) with filters applied, TCS Audio TM112.

A two-way loudspeaker from TCS Audio (TM112) comprised of a 12 inch woofer and a similarly sized horn is used to demonstrate a typical optimization process. A GLL model was created with the measured directional response data of each individual source. This model was then used to apply crossover and equalization filters to these sources.

The initial configuration of the system consisted of a fourth-order Linkwitz-Riley low-pass and high-pass filter setting at 1.6 kHz for the woofer and for the horn, respectively. Its on-axis frequency response is shown in Fig. 2.37a. Since the horn is located above the woofer, the vertical directional response can be optimized, such as for high uniformity and smoothness. A look at the 6-dB beamwidth plot of the initial setup reveals a significant narrowing of the vertical opening angle between 1 kHz and 2 kHz (Fig. 2.38a).

By taking into account the acoustical response of the woofer and the horn (both magnitude and phase), more appropriate low-pass and high-pass filter functions were selected. These new filter functions better complement the acoustic response of the transducers to yield a better overall system response as shown in Figures 2.37b and 2.38b. The new filters were a fourth-order Butterworth LP at 1.6 kHz and a fifth-order Butterworth HP at 2.0 kHz. Some additional minor equalization was also employed.

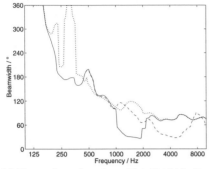
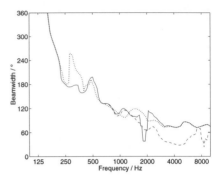

(a) Measured, configuration with initial Linkwitz-Riley filters.

(b) Predicted, configuration with optimized asymmetrical Butterworth filters.

Figure 2.38.: Vertical beamwidth of active two-way loudspeaker for woofer (- -), horn (···), and combined (−) with filters applied, TCS Audio TM112.

The vertical beamwidth through the crossover region is now much more consistent. The bandwidth over which the coverage angle decreases has been greatly narrowed; from more than one octave to approximately 1/5th octave.

Note that the beamwidth of the horn is not well defined for the lower frequencies. Similarly the beamwidth for the woofer is not defined in the high-frequency range. As a consequence the beamwidth plots show some artifacts for these regions. While beamwidth plots only yield a snapshot of the coverage, they are useful for rough comparison. More details of the directivity in the vertical plane can be seen e.g. in the vertical directivity map [100].

In order to verify the prediction, the optimized directivity filters were implemented on a readily available DSP unit. This DSP was used to drive two identical amplifier channels that powered the LF and HF sections of the loudspeaker system. The vertical directional response of the GLL model and of the measured system are shown in Fig. 2.39 for the crossover frequency of 1.9 kHz. The plot shows that there is good agreement between the prediction and the measurement.

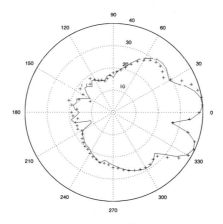

Figure 2.39.: Prediction (−) and measurement (+) of vertical directional response, configuration with optimized asymmetrical Butterworth filters, 1900 Hz at 1/3rd octave bandwidth, TCS Audio TM112.

2.4.4. Column Loudspeakers

The last part of this validation of the CDPS model is concerned with the comparison of measurement results with modeling results for two different column loudspeakers, the Alcons Audio QR36 ribbon loudspeaker and the Renkus-Heinz Iconyx IC-16 steerable array (Fig. 2.40).

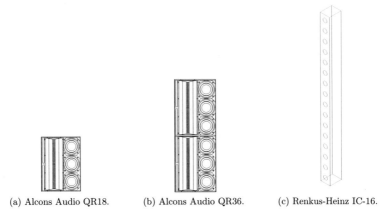

(a) Alcons Audio QR18.   (b) Alcons Audio QR36.   (c) Renkus-Heinz IC-16.

Figure 2.40.: Column loudspeakers.

Simple Column Loudspeaker

As a first example, directional measurement data are compared with the predicted performance of a simple column loudspeaker. The simulation results were obtained by modeling a stack of two Alcons Audio QR16 loudspeakers (Figs. 2.40a, 2.40b) on the basis of a single full-range directional data set. This stacked loudspeaker setup is physically identical to a QR36 loudspeaker which was measured for comparison.

Figure 2.41 shows the vertical directional response of the simulated and measured QR36 system at different frequencies. Clearly, there is a good match especially for the front half. However, at higher frequencies, such as 10 kHz (Fig. 2.41d), a mismatch becomes apparent. It is caused by the insufficient angular resolution of the 5° measurement data that is the basis for the model (see Sections 2.4.2 and 3.2.3). This aspect is particularly important for ribbon loudspeakers like the QR18 and QR36 which are normally highly directed in the high-frequency range.

This example shows that a set of acoustic sources or drivers (in this case four) can be measured as a single data set and then used to describe the radiation characteristics of a system that utilizes multiple such data sets. Especially in this case, the QR36 system with a height of about 1 m reaches the physical limitations of any 3D full-sphere measurement system for loudspeakers available today. It may even be possible to handle

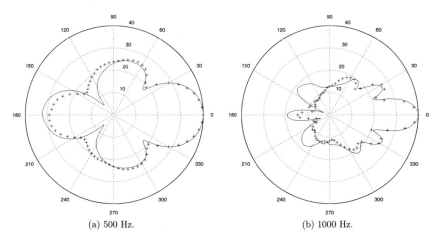

(a) 500 Hz.

(b) 1000 Hz.

Figure 2.41.: Prediction using two large full-range sources (−) and measurement (+), complex model, vertical directional response, 1/3 octave bandwidth, Alcons Audio QR36.

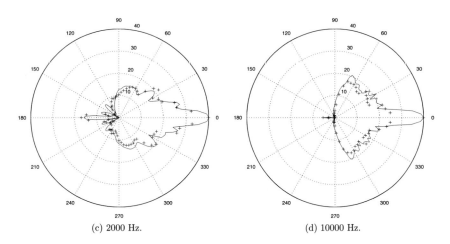

(c) 2000 Hz.

(d) 10000 Hz.

Figure 2.41.: *Continued.*

boxes of greater extension mechanically. However, the measuring distance becomes too short compared to the loudspeaker size in order to be still approximately in the far field.

Digitally Steered Column

The second example is a column loudspeaker that is digitally steerable by means of an embedded DSP controller. The loudspeaker is 2 m tall and consists of 16 separate output channels each driving a set of of three transducers, namely a cone driver and two dome-tweeters mounted on top of it. The triplets are equally spaced along the height of the array, see Fig. 2.40c. Due to the loudspeaker's extent, both measurements and simulations were performed in the near field of the device, at approximately 6.5 m. For the comparison a DSP setting was used that represents a nominal opening angle of 20° without inclination of the beam. The effective origin of the beam was chosen to be in the middle of the array. The system was modeled on the basis of the raw measurement data for one triplet of transducers and separate DSP filter settings for each channel. It was then compared with measurement results for the column using the same filters. This setup is obviously similar to the foregoing Section 2.4.3, but in this case multiple full-range sources are equipped with filters and not a crossover system.

Figure 2.42 shows the measured and simulated vertical directional response for selected frequencies. The displayed frequencies depict the polar characteristics below the aliasing frequency (2.42a), close to it (2.42b) and above it (2.42c). There is a good match of the main lobe and the first side lobes which continues even up to 10 kHz (2.42d).

In the area of 60° to 120° off-axis for the mid-frequency range deviations of up to about 10 dB are visible. The vertical asymmetry of the measurements for the symmetric loudspeaker indicates measurement errors. In fact, these are largely artifacts due to high-level reflections at the measurement robot. Even though some absorption materials were in place, these artifacts could not be suppressed entirely. Additionally, there is only a single main lobe in the frequency range considered, and the directional data are normalized to this peak. This is an error-sensitive situation, because when the on-axis measurement does not exactly capture the maximum of the main lobe, the level for other, off-axis angles is increased artificially due to the normalization to this peak (see also Chapter 3). In any case it should be noted that the levels discussed are on average 20 dB or more below the main lobe.

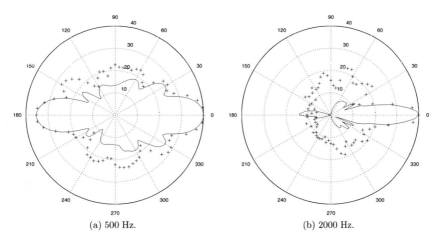

(a) 500 Hz.　　　　　　　　　(b) 2000 Hz.

Figure 2.42.: Prediction using 16 full-range sources with DSP filters (−) and measurement (+), complex model, vertical directional response, 1/3 octave bandwidth, Renkus-Heinz IC-16.

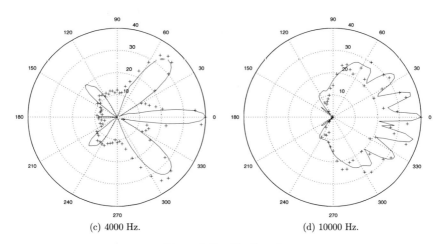

(c) 4000 Hz.　　　　　　　　　(d) 10000 Hz.

Figure 2.42.: *Continued.*

2.5. Summary and Open Problems

It was shown that when combinations of sources are to be simulated the use of complex-valued directional data for describing the radiation characteristics of sound sources is significantly more precise than using only magnitude data. Magnitude-only data may yield satisfying results when each acoustic source or transducer is measured and modeled individually. It is a pre-condition, however, that for each source the inherent phase response can be neglected and the exact origin of the radiated sound wave is determined in 3D. Both requirements represent severe limitations for practical application.

The CDPS model resolves a number of problems that were connected with table-based or magnitude-only data. But it should be emphasized that this model is based entirely on the assumption of point sources in the high-frequency limit. Boundary and diffraction effects by the loudspeaker enclosure or adjacent boxes can only be accounted for indirectly, namely by including them in the directional measurements. Also, in its description of loudspeakers as electro-acoustic devices the practical implementation of the CDPS model by means of the GLL concept has several shortcomings, still. These include in particular:

- Although filter functions are fully supported, electronic coupling among components of a loudspeaker, such as between a passive filter network and a transducer, are not. This requires a two-port modeling approach which is essentially only important for passive loudspeakers.

- The maximum sound pressure level for each source is determined by its sensitivity and a given maximum input voltage. But in practice the latter quantity is not well defined with respect to its actual criterion. Additionally, slightly below and everywhere above that line, the loudspeaker still radiates sound but with a rather nonlinear response behavior.

- Another aspect worth mentioning is the treatment of uncertainties regarding the actual directional data. This applies to both, the measurement process as well as the variation among different production samples of the same loudspeaker model. The issue of sample variation will be addressed to some extent in Chapter 3.

3. Modeling Large Sound Sources

The concern of the previous chapter has been *small sound sources* primarily. They have been defined as acoustic sources that are used mainly in their far field and that can be considered as point sources. In this chapter *large sound sources* will be discussed which cannot be approximated by a single point source because they are used into their near field. Spatially extended sound sources from the electro-acoustic domain [38], [39], [40], [41] will be considered, but not physically large sources that radiate sound in an incoherent manner, such as highways with heavy traffic. Typical representatives from the field of sound reinforcement include:

- Touring or concert line arrays consisting of a large amount of cabinets that are mounted below each other at different splay angles (see e.g. Figs. 1.3, 3.10, 3.22),

- Large column loudspeakers which consist of many small transducers with or without digital control (see e.g. Fig. 2.40),

- Loudspeaker clusters which consist of an array of loudspeakers that are usually arranged in a matrix-like form in both the vertical and horizontal domain,

- Piston-like loudspeakers which have a rectangular or circular shape and consist of many small elemental loudspeakers,

- Loudspeakers that consist of one or multiple ribbon transducers which provide highly directional radiation patterns.

As diverse as this selection may seem, all of these systems have in common that their purpose is not only increased output sound power but also enhanced directional control. For these reasons, such systems are relatively large compared to conventional loudspeakers; usually they are several meters tall. Their design is intended to optimize the interaction between individual elements in order to minimize destructive interference effects and generate a coherent wave front.

It will now be discussed how these acoustic sources can be modeled, with a focus on the example of a finite line source. In the next section existing practices will be reviewed briefly. After that a new, more advanced approach called *CDPS decomposition* will be introduced and its accuracy demonstrated [50]. Finally the effect of production variation of loudspeakers on the performance of an array of loudspeakers and the corresponding model will be examined.

3.1. Conventional Source Models

It is immediately clear that the more complicated the design and the control of a sound source or loudspeaker system, the greater the need for appropriate modeling facilities. Accordingly, a variety of different approaches have been introduced in the past in order to describe the aforementioned types of sources numerically for the purpose of acoustic simulation, e.g. [103], [104], [105], [106], [107], [108], [109], [53]. The most important methods will be outlined in the following sections.

All of these models share some basic concepts. They try to simulate both near field and far field of comparably large sound sources. Also, they typically use measurement data that are limited to magnitude data and that are acquired at significantly fewer points than a regular directional data set. This type of approach was maybe most feasible ten or fifteen years ago when the acquisition of reliable phase data was still considered complicated and when performing full-sphere directional measurements was relatively difficult. However, there are inherent drawbacks to these simplified methods that cannot be overcome easily. These drawbacks have become a dominant factor now that more advanced measurement and modeling methods are available.

For a start, the simple example of a straight, finite line source should be used. Most of the algorithms mentioned above try to model linear or curved sources in order to reproduce the wave front generated by a transducer array or a wave guide. But basically for any physically large source there is no closed analytical solution that can be used in a simple computer model. Although the analytical definition of the finite line source is known, the solution of the problem is difficult. The same is true, e.g., for the circular source and the rectangular source.

The finite line source is accurately determined by the integral

$$p(r, \theta, t) = A_0 \int\limits_{-L/2}^{L/2} \frac{1}{r'} e^{i(\omega t - k r')} dx, \tag{3.1}$$

where p is the complex pressure at the receiver distance r and angle θ at a time t, with r relating to the center of the line source. The integral covers the length L of the line source in elements dx where r' denotes the distance of the element to the receiver (see Fig. 3.1). The complex radiation amplitude is denoted by A_0, ω is the radial frequency and k is the wave number [1].

Such integrals can be resolved by making appropriate simplifications, such as the far-field approximation or the decomposition into elementary sources. The latter is related to numerical integration which is a viable option in some cases, as well.

3.1.1. Far-Field Model

One of the most commonly used approaches for solving problems of the kind of eq. (3.1) is to apply a far-field assumption. By definition, this approximation cannot account for near-field effects. However, it provides at least an accurate description for large receive

3. Modeling Large Sound Sources

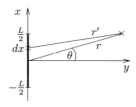

Figure 3.1.: Geometry of finite line source.

distances and a reasonable estimate for the transition region between near field and far field.

Far-Field Condition

The far field is normally considered as reached at distances where an extended source can be approximated by a point source[1]. However, only at infinite distance the point source behavior is truly assumed. Because the transition from the near field into the far field is smooth, a quantitative condition must be given that defines how large the deviation from the exact asymptotic state is allowed to be in order for a location to be still in the "approximate far field".

There are various ways to define such a condition. They normally involve the assumption that the receive distance r is large compared to the dimensions of the source, L. Also, the size of the source relative to the wavelength $\lambda = 2\pi/k$ plays a role. A typical condition is given, for example, by limiting the phase error at a finite receive distance relative to an asymptotically infinite receive distance.

Here it is assumed that this deviation can be approximated by the propagation phase difference between the closest and the furthest point of the source at a receive location relative to infinite distance. These three points span a triangle whose sides are established by the connection line D between the closest and the furthest point, the shorter leg R, and the longer leg $p + q$ (see Figs. 3.2a and 3.2b).

Accordingly, at the receiver location the difference Δ_R between the path lengths from the two source points is given by:

$$\Delta_R = p + q - R. \tag{3.2}$$

Relative to the far field, where $R \equiv q$ and thus the path length difference $\Delta_\infty = p$, one obtains a propagation phase difference of

$$\Delta\phi = \frac{2\pi}{\lambda}(\Delta_\infty - \Delta_R) = \frac{2\pi}{\lambda}(R - q). \tag{3.3}$$

---

[1]One also speaks about the property that the connecting lines between different points of the extended source and the receiver are considered *parallel* in the far field. In optics this is known as the *Fraunhofer condition* [110].

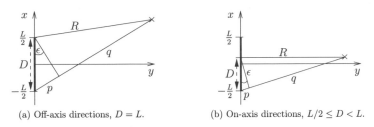

(a) Off-axis directions, $D = L$. (b) On-axis directions, $L/2 \leq D < L$.

Figure 3.2.: Far-field approximation of finite line source.

Introducing the angle ϵ between the line through the two source points and the perpendicular of the long side, $p + q$, yields

$$\Delta\phi = \frac{2\pi}{\lambda}(R - \sqrt{R^2 - D^2 \cos^2 \epsilon}). \tag{3.4}$$

Since one can assume that $D \ll R$, only first-order terms may be kept and one finds:

$$\Delta\phi \approx \frac{\pi D^2 \cos^2 \epsilon}{\lambda R}. \tag{3.5}$$

For a given phase error $\Delta\phi$ this yields a condition for the nominal far-field radius R_{Far}:

$$R_{Far} > \frac{\pi D^2 \cos^2 \epsilon}{\lambda \Delta\phi}. \tag{3.6}$$

Looking for a simple lower-bound estimate of R_{Far} one may conservatively use $\cos^2 \epsilon \leq 1$ so that:

$$R_{Far} > \frac{\pi D^2}{\lambda \Delta\phi}. \tag{3.7}$$

Assuming that the phase error should not exceed π, the receiver location must satisfy:

$$R_{Far} > \frac{D^2}{\lambda}. \tag{3.8}$$

Note that for small angles ϵ, in directions close to the radiation axis, the closest source point is the perpendicular projection of the receiver point on the line source (Fig. 3.2b). This equates to $L/2 \leq D < L$ when $\epsilon < \arctan(\frac{L}{R})$ and $D = L$ when $\epsilon \geq \arctan(\frac{L}{R})$. For a different shape of the extended source, D must be determined accordingly.

For the on-axis direction with $\epsilon = \arctan(\frac{L}{2R})$ one has $D = L/2$. For a phase error of π, the receiver location must therefore satisfy:

$$R_{Far} > \frac{L^2}{4\lambda}. \tag{3.9}$$

For off-axis directions $D = L$ applies, so that

$$R_{Far} > \frac{L^2}{\lambda}. \tag{3.10}$$

In many electro-acoustic applications it has become common to set the far-field radius to

$$R_{Far} > \frac{L^2}{2\lambda}. \tag{3.11}$$

To conclude, the derivation of the far-field condition was demonstrated only for the straight line source but the method can be easily generalized to any spatially extended source.

Far-Field of the Finite Line Source

Eq. (3.1) can be solved for the ideal far field [1]. This assumption means that the denominator can be approximated for large distances $r' \approx r$ so that it does not depend anymore on the integration variable x. The phase in the exponent must be treated more carefully since the term may depend strongly on the angle. Therefore the first order correction has to be included, $r' \approx r - x \sin\theta$, which leads to

$$p(r, \theta, t) = \frac{A_0}{r} e^{i(\omega t - kr)} \int\limits_{-L/2}^{L/2} e^{ikx \sin\theta} dx. \tag{3.12}$$

This integral can be solved and one obtains:

$$p(r, \theta, t) = \frac{A_0 L}{r} \frac{\sin(\frac{1}{2}kL \sin\theta)}{\frac{1}{2}kL \sin\theta} e^{i(\omega t - kr)}. \tag{3.13}$$

Obviously, this far-field solution is given by a simple point source with a directional term that can be considered as a sinc function over the angle θ.

3.1.2. Far-Field Cluster of Directional Point Sources

A topic that is related to the far-field approach for a single, large source is the combination of multiple point sources in a so-called *cluster* [104], [105]. In this case it is defined as a single, virtual sound source that reproduces the far-field properties of the loudspeaker arrangement. The directional response of the cluster can be determined by accounting for the directional response of each point source as well as for the offset of each source relative to the other point sources. In particular for applications where the zones of interest are located in the far field of the entire cluster itself, this method can reduce the overall calculation effort significantly. That is because the far-field directional response only needs to be computed once for each cluster configuration and can then be used subsequently for many mapping points, calculation runs, and room models.

For many years, this approach was the only practical way of modeling large sound systems. Until roughly ten years ago it was impossible or at least difficult to treat the individual point sources throughout a system modeling process including detailed numerical computations, such as ray-tracing. It was similarly impractical to measure the directional function of the cluster as a whole.

However, there are some general disadvantages to this approach that should be mentioned shortly. Naturally, the configurability of loudspeaker clusters cannot be accounted for without re-determining the far-field directional data after each change of the setup. When the optimization of the setup takes many iterations, this additional calculation step may cause a significant overhead. Also, by definition the validity of the approach is limited to the far field of the system and does not allow for realistic near-field results.

Finally, depending on the available directional data there may be aliasing problems [111]. The resolution of the measurement data for the individual sources must be high enough in order to allow for meaningful interpolation since the spatial data points of the resulting cluster data do not coincide normally with the data points of the included sources. In many cases, accurate results will necessitate acquiring phase data for the included sources, too.

Given the increase in computing power and memory over the last decade the reduction of the loudspeaker array to a single source is not necessary anymore. Rather, the sum of point sources can be used directly. One can use the time-independent version of the propagation equation (1.4) to calculate the overall sound pressure produced by an array at the receiving location based on the combination of the individual elements j of the array,

$$p(\vec{r}, \omega) = \sum_{j=1}^{N} p_j(\vec{r}, \omega), \tag{3.14}$$

where each summand p_j represents a particular loudspeaker in the array of N elements. This pressure function p_j would normally be given by eq. (1.9), so that

$$p(\vec{r}, \omega) = \sum_{j=1}^{N} \frac{g(\hat{A}_j)}{|\vec{r} - \vec{r}_j|} e^{-ik|\vec{r} - \vec{r}_j|}, \tag{3.15}$$

where $\hat{A}_j$ is the angle- and frequency-dependent complex radiation function.

However, like the cluster approach, in practice this method was limited by the fact that loudspeaker phase data were normally not accounted for which means $\hat{A}_j \equiv |\hat{A}_j|$ and $g \equiv |g|$. In most cases only the phase due to the propagation delay, $k|\vec{r} - \vec{r}_j|$, was taken into account for each source when calculating the complex sum. The same issue was discussed in Chapters 1.3 and 2 in the context of modeling small sound sources. Eventually, the inaccuracy of these results led to a number of derived or alternate approaches some of which will be introduced in the next sections.

3.1.3. Numerical Solution

A relatively simple method of solving problems, e.g., as given by eq. (3.1), is the brute-force numerical solution of the source integral. In this case the continuous integral is approximated by the discrete sum

$$p(r, \theta, t) = A_0 \frac{L}{N} \sum_{j=1}^{N} \frac{1}{r_j'} e^{i(\omega t - k r_j')}. \tag{3.16}$$

This representation has an interesting practical interpretation. The number of summands corresponds directly to the amount of imagined elementary sources j arranged in an array. The spacing L/N equates to a spatial resolution which defines the upper frequency limit for the validity of the numerical solution.

However, this solution is only applicable when the curvature and location of the source are known and well defined. In practice, many loudspeaker systems exhibit radiation characteristics similar to e.g. a finite line source but the origin of the propagating wave may be something different than a line source. This has led to methods that try to reproduce the resulting wave front instead of the actual source of the sound wave.

Elementary Sources Approach

The Huygens principle [112], [113] states that a propagating wavefront can be reproduced by a set of discrete, omnidirectional point sources located on that wavefront, e.g. similar to eq. (3.16). Various attempts have been made to employ Huygens' principle as a computational radiation model for curved or linear sound sources; see, for example, [106], [107].

In its simplest form this concept is only applied in the vertical domain. The radiation from the mouth of a horn or waveguide is measured at a specific distance and then modeled by a linear array of omnidirectional point sources, which are located close to each other compared to the wavelength and which represent a line of constant phase. More advanced versions allow curved or arbitrarily shaped vertical arrays of directional sources. Beyond the purely vertical domain this Huygens model can be extended to the horizontal domain as well by utilizing two-dimensional arrays of sources to represent two-dimensional sound-radiating surfaces.

All of these approaches need to be combined with an extensive set of actual measurement data. The more degrees of freedom are available regarding source count, directionality, and placement, the more measurements are needed to find optimal reproduction parameters. Due to the complexity of the optimization process and its convergence for the more advanced models, the calibration of such a source array can be a tedious procedure.

It is an advantage of this approach that no phase data have to be acquired directly, especially compared to the CDPS model, introduced earlier. It is also sometimes beneficial that a well-defined Huygens model is able to reproduce the near field of the radiating source where the CDPS model relies on far-field data.

On the other hand, the computational efforts for the directivity prediction are much higher. Considering a resolution requirement of a half-wavelength, to be valid up to 10 kHz a vertical model of a 0.3 m line source has to include at least 18 sources. Compared to the CDPS model of the cabinet, possibly using just a single source, necessary calculation times are higher by an order of magnitude. Extensions to the horizontal domain further increase the computational demands.

Another disadvantage of the Huygens approach is its lack of generality regarding off-axis radiation angles. Naturally the wavefront principle cannot be applied when planes of constant phase are difficult to measure or do not exist. This is true especially for the sides and the back side of a typical loudspeaker cabinet. In addition to that, wavefront models that consist of only a vertical array of sources need to approximate the pressure radiation for the horizontal domain. This is often accomplished by including a conventional directivity measurement of the horizontal plane. However, it is questionable and remains to be proven that for the diagonal planes there exist interpolation methods of satisfying accuracy.

3.1.4. Integral Methods

A very accurate solution to the problem of determining the acoustic field exterior to an object with radiating elements exists in the form of BEM [21]. Unfortunately an application of the method to a complete array over the entire audible frequency range poses a substantial computational difficulty. Largely for this reason researchers have concentrated on utilizing BEM only for individual radiating components within a single array element.

In an effort to reduce the scale of the problem some attempts have been made to apply Rayleigh integral techniques [114] to characterize a radiating component within an array. The complex pressure or normal velocity is deduced over a small flat surface in front of the component. Once known, the exterior field can be determined in front of the plane. The idea is then to tessellate these surfaces to simulate an array. Leaving aside the inability of the approach to model the rearward radiation there remains the problem of acquiring the surface data. One method is to make some polar measurements of the device and use them as a target in an optimization loop that attempts to determine the surface data [108]. Direct measurement of the data may also be possible [109]. Alternatively, for horns one could view the problem as an interior BEM solution, with appropriate boundary conditions on the imagined surface, thus obtaining surface data based on the real horn geometry [53].

Assuming accurate surface data for one component to be available, there still remains the problem of the influence of the rest of the array. To illustrate this, a mid-frequency horn in a medium-sized touring line array has been modeled using BEM. Fig. 3.3a shows the 400-Hz pressure magnitude on an imagined surface just behind the grille. The surface extends over a six-box array with the active box at second position from the top. Fig. 3.3b shows the effect of simply placing inactive elements above and below the active one. Obviously one will obtain quite different results for the exterior fields for these two configurations at this frequency.

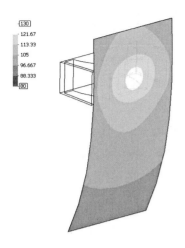 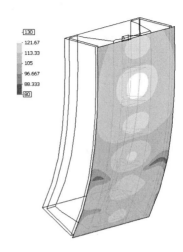

(a) Imagined surface in front of isolated loud-speaker.

(b) Imagined surface in front of an array of six loudspeakers with only the second box being active.

Figure 3.3.: Modeled pressure magnitude at frequency of 400 Hz. Light colors indicate zones of high pressure magnitude, dark colors represent zones of low pressure magnitude.

The integral approaches can be useful for designing components or modeling isolated loudspeakers. However, when applied to wide-band array predictions they have similar limitations as the CDPS model of a single loudspeaker cabinet. Many of the concerns these methods address, such as geometric error mechanisms, are solved by the use of complex data in the CDPS model. So it seems that the considerable measurement or analysis burden to create such integral models is not warranted. As will be shown later in this chapter, a combination of a CDPS model operating in the region of geometrical acoustics and BEM operating below that frequency could be considered ideal.

3.2. CDPS Decomposition

In the preceding sections the fundamentals of ideal, finite line sources have been discussed. In reality the approximation of an ideal line source, the so-called line array, finds widespread use. Several loudspeaker cabinets are combined to form a linear or even curved array of sources. Individual cabinets may reproduce the behavior of an ideal line source depending on the design, the frequency range considered, and the intended use.

3. Modeling Large Sound Sources

It is quite obvious that if a loudspeaker can be approximated by a point source according to equations (1.7) and (1.9), an array of such loudspeakers will - following the fundamentals of linear superposition - show the radiation behavior as described by eq. (3.15), assuming the effects of neighboring cabinets on each other can be incorporated.

In practice many arrays of loudspeakers are claimed to be close to the mathematically ideal line source, which means that over the usable frequency range of the device the array acts approximately like a single line source. It will now be shown that any real-world line source can be described reasonably by the CDPS model.

3.2.1. Decomposition

Relationship (3.1) can be rewritten identically to a sum of partial integrals, each representing, for example, a cabinet of the line array,

$$p(r,\theta,t) = A_0 \sum_{j=1}^{N} \int_{-L/2+(j-1)L/N}^{-L/2+jL/N} \frac{1}{r'} e^{i(\omega t - kr')} dx. \tag{3.17}$$

As a second step one can apply a coordinate transformation that places the origin central to the boundaries of each partial integral, $x_j = x - [-L/2 + (j-1)L/N]$. The corresponding transformations are applied to $r(x)$, $\theta(x)$, and $r'(x)$ in the same manner. This yields

$$p(r,\theta,t) = A_0 \sum_{j=1}^{N} \int_{-l/2}^{l/2} \frac{1}{r'_j} e^{i(\omega t - kr'_j)} dx_j. \tag{3.18}$$

Here the subscript j denotes the dependence on the particular coordinate system. The parameter $l = L/N$ represents the length of the individual element, which is identical to the spacing between centers of adjacent elements.

For each partial integral one can make the far-field assumption that $r'_j \gg l$ so that $1/r'_j \approx 1/r_j$ and $\exp(-ikr'_j) \approx \exp[-ik(r_j - x_j \sin\theta_j)]$. Also one is free to rename the integration variable x_j as x,

$$p(r,\theta,t) = A_0 \sum_{j=1}^{N} \frac{1}{r_j} e^{i(\omega t - kr_j)} \int_{-l/2}^{l/2} e^{ikx \sin\theta_j} dx. \tag{3.19}$$

After that the partial integrals can be solved and partial solutions are obtained in the form of eq. (3.13):

$$p(r,\theta,t) = A_0 l \sum_{j=1}^{N} \frac{1}{r_j} e^{i(\omega t - kr_j)} \frac{\sin(\frac{1}{2}kl\sin\theta_j)}{\frac{1}{2}kl\sin\theta_j}. \tag{3.20}$$

Obviously this sum can be understood as a set of point sources with the directivity function

$$\Gamma(\theta) = \frac{\sin(\frac{1}{2}kl\sin\theta)}{\frac{1}{2}kl\sin\theta}. \tag{3.21}$$

82

This proves that eq. (3.15) also applies when a continuous line source is subdivided into smaller line sources. But it is only valid in the far field of the partial source. Similar to the numerical solution (3.16) of the line-source integral there is a resolution, l, that determines the accuracy of the approximation and the computational effort simultaneously. Compared to the numerical solution, the additional directional factor in eq. (3.20) can be understood as a correction to the omnidirectional elementary source that allows for a wider spacing of the elements.

The concept of subdividing the given integral and solving for the far field of the individual element is one of the key results of this thesis. Note that the above derivation does not incur any loss of generality. It is fully applicable to any continuous source, such as curved lines.

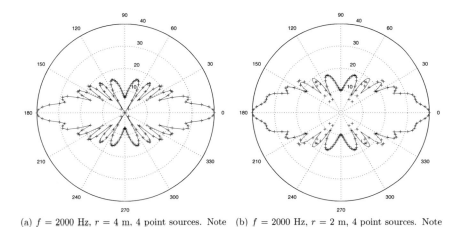

(a) $f = 2000$ Hz, $r = 4$ m, 4 point sources. Note that spacing $l = 0.25$ m is larger than wavelength $\lambda = 0.17$ m.

(b) $f = 2000$ Hz, $r = 2$ m, 4 point sources. Note that at 2 m one is entering the near-field zone of the individual source.

Figure 3.4.: Comparison of line source integral $(-)$ with approximation by point sources with complex directivity $(+)$. Length of the line source is $L = 1$ m.

3.2.2. Examples

The four directional plots of Fig. 3.4 demonstrate the applicability of the CDPS decomposition using a few selected examples. All of them show vertical polar plots of the radiation of an ideal (vertical) line source at a radial scale of 40 dB. In all cases the solution provided by the CDPS model, eq. (3.20), is compared with the accurate numerical solution, eq. (3.16), of the original integral, eq. (3.1).

Fig. 3.4a shows the simulation of a continuous source of 1 m length by four point sources with complex directivity data at a measurement distance of 4 m. Although the

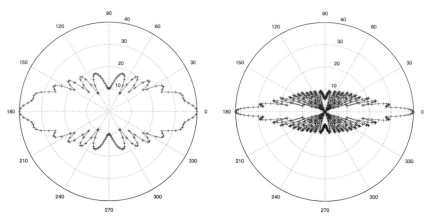

(c) $f = 2000$ Hz, $r = 2$ m, 8 point sources. At 2 m an element size $l = 0.125$ m is sufficient.

(d) $f = 4000$ Hz, $r = 8$ m, 4 point sources. Width and size of the lobes are well reproduced with an element spacing $l = 0.25$ m, even at wavelength $\lambda = 0.085$ m.

Figure 3.4.: *Continued.*

spacing of 0.25 m between the elements is significantly larger than the half-wavelength at 2 kHz, the match is very good. Fig. 3.4b shows the same setup at a measurement distance of 2 m. At this distance the measurement point begins to move into the near field of the individual element. Accordingly small deviations begin to appear. In contrast, Fig. 3.4c shows the same configuration and distance but now using eight elements with a size/spacing of 0.125 m. Obviously an exact match is re-established because the far-field condition is fulfilled again for the decomposition elements.

Finally Fig. 3.4d shows a different setup, namely, the comparison of four elementary sources using complex directivity data with the exact solution at a frequency of 4 kHz and a distance of 8 m. As can be seen clearly, main lobe and substructures match very well although the element spacing is almost an order of magnitude larger than half the wavelength $\lambda/2$. The match continues further up to higher frequencies but due to the density of side lobes this is graphically insufficiently resolved in a polar plot.

Fig. 3.5 displays the same data in a different format. The mean deviation of the CDPS approximation from the exact integral is shown as a function of frequency. It was computed for a fixed source length $L = 1$ m as the average of the absolute deviation in dB over the vertical angle θ. Fig. 3.5a compares different measurement distances r and Fig. 3.5b compares different amounts of elementary sources N. The mean deviation is greater for higher frequencies, shorter measurement distances and fewer elementary sources. Note that the mean deviation should be understood as an indicator only. It is dominated by deviations at angular points close to the nulls of eq. (3.20) where

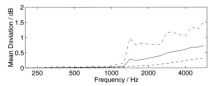 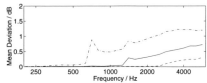

(a) Using 4 point sources, at distances of $r = 2$ m (-.-), $r = 4$ m (−) and $r = 8$ m (- -), 1/6th octave resolution.

(b) At a distance of $r = 4$ m, using 2 point sources (-.-), 4 point sources (−) and 8 point sources (- -), 1/6th octave resolution.

Figure 3.5.: Mean deviation of line source integral and approximation by point sources with complex directivity. Length of the line source is $L = 1$ m.

the directivity function has steep slopes. This is also the reason for the partly non-monotonous dependency of the mean deviation on frequency.

3.2.3. Data Requirements

In order to describe a line source as characterized by eqs. (3.13) or (3.20) by the discretized CDPS model, eq. (1.9), a set of conditions must be fulfilled. These originate in the specific properties of the line source, but similar requirements are found for the far-field solution of other types of continuous sources.

Complex Data

First of all, phase data must be included. This can be seen immediately from the alternating sign of Γ in eq. (3.21) because $\Gamma(\theta) \neq |\Gamma(\theta)|$. The phase over angle θ or frequency f is basically a step function that switches between two states, 0 and π, at every zero of the directional factor. Omitting this information will lead to erroneous results when computing the linear superposition of multiple sources.

This is illustrated by Fig. 3.6, which shows the magnitude and phase of the vertical polar response for a line source of 1 m length at 2 kHz in the far field. Note that the phase was compensated for the propagation delay and normalized so that a phase value of 180° is plotted at the polar origin, and 0° corresponds to 38 dB.

Angular Resolution

Equation (3.21) also gives an idea about the angular resolution required for the discrete point source model. Based on the directional factor one can derive the angular spacing of the off-axis nulls, where $\Gamma(\theta) = 0$. From there it is only a small step to define the angular resolution needed for an accurate description by the directivity matrix $\hat{A}$.

The nulls of the angle-depending directional factor are distributed according to

$$\theta_i = \arcsin\left(\pm i\frac{c}{fl}\right), \quad i = 1, 2, \ldots. \tag{3.22}$$

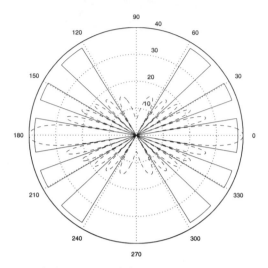

Figure 3.6.: Magnitude (- -) and phase (−) of an ideal, finite line source of length $L =$ 1 m, at $f = 2$ kHz, in the far field. Phase was scaled so that 38 dB is equivalent to $0°$. Note that phase switches at every minimum.

Here θ_i denotes every angular location i where the directional factor vanishes. As an example, for a source length $l = 0.2$ m, a frequency $f = 10$ kHz, and a speed of sound c $= 340$ m/s the first null occurs at $\theta_1 \approx 10°$, the second at $\theta_2 \approx 20°$. For shorter sources the spacing of nulls becomes wider for the same frequency (Table 3.1).

Length l/m	0.1	0.2	0.4	0.8
θ_1	19.9°	9.8°	4.9°	2.4°
θ_2	42.8°	19.9°	9.8°	4.9°
θ_3	—	30.7°	14.8°	7.3°
θ_4	—	42.8°	19.9°	9.8°
θ_5	—	58.2°	25.2°	12.3°

Table 3.1.: Angular locations of magnitude minima 1-5 for different source lengths l at frequency $f = 10$ kHz.

In order to avoid aliasing or undersampling errors, the angular resolution must be high enough given the length l of the source and the upper frequency limit. This resolution limit can be approximated by half[2] the angular distance between the first and second

---

[2]The choice of half the spacing between adjacent magnitude minima as a condition seems natural but is somewhat arbitrary. Of course, higher requirements will lead to smaller quantitative errors.

magnitude minimum,

$$\Delta\theta_{crit} = \frac{1}{2}\left[\arcsin\left(2\frac{c}{fl}\right) - \arcsin\left(\frac{c}{fl}\right)\right]. \tag{3.23}$$

For this example an angular resolution of 5° could be just sufficient. In general, angular resolutions of $\Delta\theta \leq \Delta\theta_{crit}$ should be used (Table 3.2).

Indeed, in practice another issue is equally important. Because the main lobe (on axis, $\theta = 0°$) becomes very tight for high frequencies, the on-axis data must be captured carefully in order to obtain the actual on-axis level. This is especially important for turntable measurements. Table 3.3 shows some exemplary data calculated from eq. (3.21) for angles θ close to 0°. A small angular deviation from the exact on-axis direction will result in a lower on-axis level measurement, because the pressure function changes quickly with the angle. Failing to measure the correct level of the maximum that is used for normalizing the rest of the measurements will increase the level of the sidelobes artificially and thus lead to erroneous results later on.

Length l/m	0.1	0.2	0.4	0.8
$\Delta\theta_{crit}$	10°	5°	2.5°	1°

Table 3.2.: Minimum required angular resolutions for different source lengths l at frequency $f = 10$ kHz.

Length l/m	0.1	0.2	0.4	0.8
Level/dB at 1°	-0.04	-0.15	-0.61	-2.56
Level/dB at 2°	-0.15	-0.61	-2.56	-13.70
Level/dB at 3°	-0.34	-1.40	-6.32	—

Table 3.3.: Attenuation at off-axis angles θ compared to on-axis for different source lengths l at frequency $f = 10$ kHz.

Frequency Resolution

Similar to the angular resolution one can derive from eq. (3.21) a condition for the needed spectral resolution. In the frequency domain the spacing of nulls corresponds to

$$f_i = i\frac{c}{l\sin\theta}, \quad i = 1, 2, \ldots. \tag{3.24}$$

where f_i denotes the frequencies i for which the directional factor becomes zero. For the example of $l = 0.2$ m and $c = 340$ m/s the frequency nulls occur at a spacing of 1700 Hz when considering an angle of $\theta = 90°$. The spacing is larger for smaller angles and for smaller source lengths. Table 3.4 shows some exemplary data.

3. *Modeling Large Sound Sources*

Length l/m	0.1	0.2	0.4	0.8
f_1/Hz	3400	1700	850	425
f_2/Hz	6800	3400	1700	850
f_3/Hz	10 200	5100	2550	1275
f_4/Hz	13 600	6800	3400	1700
f_5/Hz	17 000	8500	4250	2125

Table 3.4.: Frequencies of magnitude minima 1 to 5 for different source lengths l at angle $\theta = 90°$.

Because of these characteristics the frequency resolution has an upper bound. In order to resolve the spectral structure one can define the condition

$$\Delta f_{crit} = \frac{c}{2l\sin\theta},\qquad(3.25)$$

which means that in order to avoid aliasing problems there should be at least two data points for every frequency section enclosed by adjacent zero points of the directional factor. This corresponds to minimum required frequency resolutions $\Delta f \leq \Delta f_{crit}$.

The highest resolution requirement occurs for $\theta = 90°$,

$$\Delta f_{crit} = \frac{c}{2l},\qquad(3.26)$$

which equates to 170 Hz for a 1-m line source. In this respect it should be noted that practically all modern FFT-based measurement systems [115] provide frequency resolutions that are much higher than this.

Frequency Averaging

In acoustics the frequency resolution is often based on fractional octave bands, thereby reflecting the way of human sound perception. Related band averages are used to describe acoustic quantities in physiologically relevant frequency resolution limits, or to compress measured or simulated data without loss of significance. It is important to understand when it is reasonable to average data such as the modulus of the directional factor over a bandwidth.

One can imagine that the average over a frequency range that includes less than half the spectral distance between two nulls, eq. (3.24), can be used as a representative value. That is possible because the average will not depend much on the actual limits of the averaging bandwidth but rather follow the underlying function smoothly. On the other hand, if the average is computed over a frequency bandwidth of at least twice the angular distance between two zero points one can assume that the average will be representative as well. The reason is that for a large enough bandwidth additional variations of the underlying function will not be significant. But for the intermediate frequency range small variations of the particular position of the averaging interval will lead to large variations of the average value, which in turn leads to spurious results. This is the

88

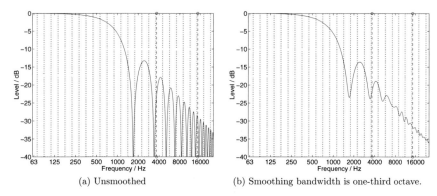

(a) Unsmoothed   (b) Smoothing bandwidth is one-third octave.

Figure 3.7.: Frequency response $(-)$ of ideal line source of length $l = 0.2$ m at an angle of 90° off-axis in the far field. Bands of one third-octave width $(\cdots)$, forbidden range from 3670 to 14 700 Hz (- -).

"forbidden" frequency range. Figs. 3.7a and 3.7b show examples for a bandwidth of one-third octave.

Given a bandwidth b in fractional octaves one can calculate the exact center frequency of the related band that has the width of the linear frequency resolution Δf,

$$f_c = \frac{\Delta f}{2^{b/2} - 2^{-b/2}}. \tag{3.27}$$

Based on this one can determine directly for which frequencies f_c and fractional octave resolutions b averaging is allowed. For the lower frequency limit the linear resolution in eq. (3.27) is given by half the spacing of magnitude minima according to eq. (3.25),

$$f_{lower} = \frac{c}{2l \sin \theta} \frac{1}{2^{b/2} - 2^{-b/2}}. \tag{3.28}$$

The upper frequency limit is given by twice the spectral spacing between magnitude minima,

$$f_{upper} = \frac{2c}{l \sin \theta} \frac{1}{2^{b/2} - 2^{-b/2}} = 4 f_{lower}. \tag{3.29}$$

In the intermediate range of $f_{lower} < f_c < f_{upper}$ the frequency data points are too coarse and will generate aliasing errors due to the quickly varying average that is sampled using points that are too far apart[3].

As an example, for $l = 0.2$ m, $c = 340$ m/s, $\theta = 90°$, and a resolution of one-third-octave bandwidth $b = 1/3$, data can be averaged meaningfully below 3670 Hz and above

---

[3]The potential error can be quantified by approximating the band average of eq. (3.21) and analyzing its variations over the frequency range.

14 700 Hz (Fig. 3.7a). If the length of the source is greater, these frequency limits will be reduced. They will increase for smaller angles.

A selection of different values is shown in Table 3.5. It is obvious that the forbidden frequency range depends strongly on the angle and can thus not be understood as a fixed range for a given source length l. In general, one can derive two different regimes.

- For large sources ($l > 2$ m) averaging over wide bandwidths, such as 1/1 octave, may be useful to derive representative data in a statistical way.

- For small to medium-size sources ($l < 0.5$ m) averaging over small bandwidths, such as 1/24 octave or 1/36 octave, may be used to smooth the frequency response curve without losing significant information.

For intermediate averaging bandwidths, such as one-third octave, most data will suffer from significant sampling errors. That can only be avoided for carefully selected angles and frequencies of interest. As an example, Fig. 3.7b displays the frequency response of Fig. 3.7a smoothed to one-third octave. In the forbidden range the distance between one-third-octave center frequencies is too large to capture the fundamental behavior of the average function. In this case variations of up to 6 dB occur between adjacent data points so that the reduced data set depends strongly on the actual placement of the sample points.

	$l = 0.2$ m		$l = 0.8$ m	
Bandwidth b	f_{lower}	f_{upper}	f_{lower}	f_{upper}
1/1 at $\theta = 20°$	3500	14 000	900	3500
1/3 at $\theta = 20°$	10 700	43 000	2700	10 700
1/24 at $\theta = 20°$	86 000	344 000	21 500	86 000
1/1 at $\theta = 60°$	1400	5600	350	1400
1/3 at $\theta = 60°$	4200	17 000	1100	4200
1/24 at $\theta = 60°$	34 000	136 000	8500	34 000

Table 3.5.: Forbidden frequency ranges, in Hz, for different source lengths l, fractional octave bandwidths b, and angles θ.

Figs. 3.8 and 3.9 emphasize these findings on the basis of real-world measurements. They show the off-axis frequency response of a ribbon loudspeaker (Alcons Audio QR18, 0.5 m tall) at a vertical angle of 50°. In Fig. 3.8 magnitude minima are indicated by cursor lines. They seemingly match with the frequency response predicted by the model of an ideal line source of this size. The spacing of nulls should be about 890 Hz at this angle. Fig. 3.9 presents a continuous one-third-octave band frequency average of the measured response. This is to show that in the forbidden range of 2 to 8 kHz data sampled in one-third-octave steps have little relevance. Over this frequency range the variations of the curve are too large and sampling errors on the order of 6 dB can occur.

It should be mentioned that similar considerations with respect to resolution and averaging also apply to loudspeakers with a crossover, which can be regarded as very small line arrays in the frequency range of the crossover.

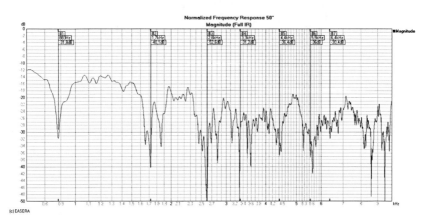

Figure 3.8.: Frequency response of Alcons Audio QR18 loudspeaker at 50° off axis normalized to 0°. Cursors denote minima of corresponding ideal line source.

Finally it needs to be emphasized that the preceding discussion regarding frequency averaging is only concerned with magnitude data. If sources are to be combined coherently, one will need to take care of phase data as well. However, especially for a line source, averaging phase data over a frequency bandwidth that contains a minimum may lead to arbitrary and meaningless results because of the switching behavior of the phase function in eq. (3.21).

Measurement Distance

With respect to the measurement distance it should be underlined that the CDPS model is only valid in the far field of the particular source. This applies to both measurement and prediction. Measurements of the sources must be taken in the corresponding far field. Calculation results will only be valid at points in the approximate far field of the modeled source (see also Figs. 3.4 and 3.5 as well as [97]). However, it should be stated once more that this is typically not the far field of the entire device. A line source may be subdivided into elements, for each of which the far-field condition must hold. But nevertheless the near field of the line source as a whole can be modeled correctly.

Errors

In this section the theoretical foundation has been laid for applying the CDPS model to spatially extended sound sources, such as line arrays. There are errors associated with that process, some of which are related to the angular and spectral resolution that is used. The minimum requirements given here seem to represent a good compromise between measurement effort and modeling accuracy in practice. Further errors may be

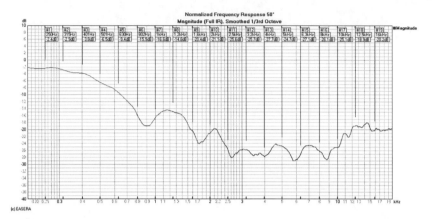

Figure 3.9.: Frequency response of Alcons Audio QR18 loudspeaker at 50° off axis normalized to 0° and smoothed at one-third-octave bandwidth. Cursors denote center frequencies of standardized one-third-octave bands.

introduced by the measurement setup, particularly with respect to complying with the far-field condition and the phase data requirements, see Chapter 2.

As mentioned in Section 3.1 an important aspect of line array modeling is the influence of the mechanical structure of the array on the radiation pattern of a single element. That will be discussed next, in Section 3.3. Another issue that plays a role when utilizing the CDPS method for line arrays are practical differences between cabinets that are assumed to be identical in theory, see Section 3.4. Finally, the effect of environmental parameters on the coherence of the elemental sound waves at the receiver will be discussed in detail in Chapter 4.

A very practical problem that will not be addressed here is related to large sound sources that cannot be subdivided for the purpose of far-field directional measurements. For example, the performance of large ribbon loudspeakers is based on the radiation of the ribbon strip that is continuous throughout the entire loudspeaker. These loudspeakers can be several meters tall, but they cannot be decomposed into partial sources so that far-field measurements are viable. It may be possible to conduct scale-model measurements or measurements in the near field of the device. These could be used to derive the directional far-field data for the entire loudspeaker or for CDPS elements.

3.3. Validation of the Model

In this section the CDPS model will be applied to typical line array systems. Validation will be based on the comparison of measurement data for the entire array with modeling results for the array based on the far-field measurements of the individual cabinets. Focus will be put on the frequency response and the directional data in the vertical domain [50]. A detailed look is taken at two different representatives of mechanically configurable line arrays, namely the Omniline by Martin Audio [116] and the SEQUENZA by Kling & Freitag [117]. Such line arrays are typically set up by adjusting splay angles between adjacent boxes. In this manner the array is curved in order to optimize the received sound pressure level throughout the listening zones.

3.3.1. Small Installation Line Array

A curved array of 12 Omniline cabinets, as depicted in Fig. 3.10, was used for this comparison. The overall vertical size of a cabinet is 0.12 m, the length of the array is about 1.4 m. All measurements were made at a distance of about 6 m which corresponds to the transition zone between near field and far field of the array and to the far field of a single cabinet.

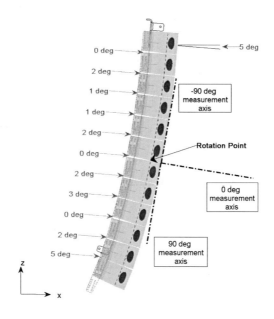

Figure 3.10.: Omniline line array setup.

In the following three models will be employed. Each of those considers the acoustic support of adjacent cabinets in a different way:

Isolated Model: First the performance of the complete system is predicted by using directional measurements of a single isolated cabinet. Although the match between measurement and prediction turns out to be already satisfying, it is clear that this method of modeling the individual point sources cannot account for mutual acoustic support and shadowing effects caused by cabinets adjacent in the array.

Flanked Model: Results can be improved when the directional data of a cabinet are acquired with the top and bottom neighbors in place. Although the two outer cabinets will be switched off and electrically short-circuited, they will contribute indirectly to a more realistic radiation behavior for the representative point source. This will be called the flanked case and it will be shown that the average deviation between measurement and simulation decreases compared to the isolated case.

An example is given in Figs. 3.11a and 3.11b, where the vertical directivity maps for the isolated and flanked cases are shown. The level attenuation relative to the on-axis direction is displayed as a function of frequency and angle.

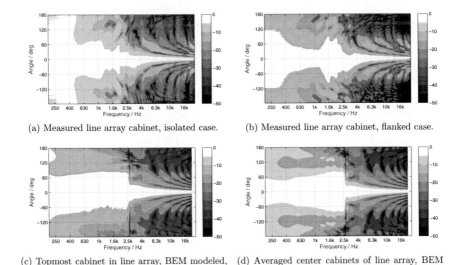

(a) Measured line array cabinet, isolated case.

(b) Measured line array cabinet, flanked case.

(c) Topmost cabinet in line array, BEM modeled, positional case.

(d) Averaged center cabinets of line array, BEM modeled, positional case.

Figure 3.11.: Vertical directivity maps. Darker colors correspond to stronger attenuation relative to the on-axis direction.

Positional Model: Consideration of the rules of superposition inevitably leads to a second step of improvement. One should think that the acoustic radiation characteristics of the outer cabinets in the array, especially the very top and bottom elements, will be different from the ones in the center. In consequence, another significant improvement can be reached by using different directional data for different cabinet positions in the array. This will be the positional model. Such data can be measured as well, although suitably accurate mechanical positioning methods for an array with a weight of possibly some hundreds of kilograms, often with the center of gravity distant from the rotation point, do hardly exist.

BEM provides a method of placing neighboring unexcited elements around the active element and acquiring complex pressure on any three-dimensional surface around that element. Previous studies had shown that when the elemental source has high enough directivity the isolated measurement is valid [118]. Also it was shown that the radiation behavior of a lower box is very close to that of a mirrored version of the corresponding upper box and that central boxes share similar radiation characteristics.

In this study elemental complex spherical data were obtained for the top four boxes of a typically curved 12-box array over the full bandwidth of the low/mid driver. An average of the centrally positioned box data was also obtained, resulting in five independent data sets. As an example, Figs. 3.11c and 3.11d show the vertical directivity maps as utilized for the topmost element and the central elements.

One could have measured isolated elemental data for the high-frequency sections of the system. However, BEM was applied to the high-frequency horn of an isolated box with a simplified driving surface. The reason, apart from the data already existing as part of the design process, was that one is freed from the usual measurement errors, such as position uncertainty and environmental factors. This advantage is particularly apparent at high frequencies.

Frequency Response

Fig. 3.12a shows the measured absolute sensitivity of the line array compared to the results predicted by the three different models. At first glance one can already see that the isolated model shows the largest deviations whereas the flanked and positional models appear to be much closer. This becomes clearer in the relative display of Fig. 3.12b, where the sensitivity data are normalized to the measurement.

For the on-axis direction the isolated case shows an average error of about ±3 dB, with maximum errors of up to 6 dB. The prediction based on the flanked setup is typically within ±2 dB, with peak errors of maximally 3 dB. The best match is reached when accounting for the position of a cabinet in the array. Here the error is about ±1 dB on average, with peaks of about 2 dB. A similar picture is given in Fig. 3.12c where the error is averaged over a set of 41 data points, namely, the relative errors within an opening angle of ±20°. The largest deviations seem to occur in three fairly separate frequency ranges.

- In the low-frequency range of about 500 Hz shadowing and acoustic support effects by neighbor elements in the array seem to be particularly dominant. Since the isolated model cannot model that, it shows the largest deviations. The positional model is better in this respect.

- In the mid-frequency range of 2-4 kHz, where the crossover is located, deviations are larger for all models. The isolated case being much worse, again due to a lack of interference from nearby irradiated surfaces, there is not a big difference between flanked and positional models. A contribution to this error likely depends on the measurement accuracy of the elements and the prediction accuracy of the array, which is normally lower in the crossover range. It will be shown later that this is the region of maximum elemental response deviation.

- In the very high-frequency range above 10 kHz deviations increase again. This is expectedly so since the measurement accuracy, and thus the prediction accuracy, becomes lower at very small wavelengths. Finite spatial precision and the influence of environmental factors during the measurement introduce noise and inaccuracy, especially into the phase data, which propagate through the prediction.

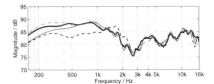 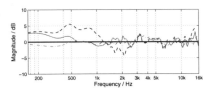

(a) Measurement (−, bold) compared to isolated (- -), flanked (−), and positional (-.-) case.

(b) Measurement (−, bold) relative to isolated (- -), flanked (−), and positional (-.-) sensitivity.

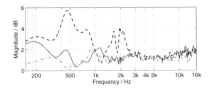

(c) Deviation of isolated (- -), flanked (−), and positional (-.-) case relative to measurement, averaged over ± 20°.

Figure 3.12.: Measured and modeled sensitivity.

Directional Response

Fig. 3.13 compares the polar responses of measured and predicted arrays for the isolated case. Similarly, Figs. 3.14 and 3.15 show overlays of the predicted vertical polar responses with measurements for the flanked and positional cases, respectively. These data were acquired from 100 Hz to 20 kHz and are displayed on graphs with a 36-dB radial scale. The angular scale denotes 0° for the on-axis direction and 90° for the upward direction. For better comparison the data were smoothed to one-sixth-octave bandwidth. Fig. 3.16 shows the same data as vertical directivity maps for an opening angle from −20° to +20°. In these plots the predicted sensitivity of the array is displayed normalized to the measurement data. Larger magnitude values, encoded as lighter colors, thus mean greater deviations.

It can be stated that all models match quite well with the measurements. Qualitatively the line array behavior is reproduced very well over the entire frequency range. Similar to the frequency response before, there is an increase in overall accuracy for the flanked model and, especially, for the positional model.

Naturally, errors are smaller for the front than for points on the back side of the array. Although still being linear, interaction effects on the back side of the cabinets cannot be predicted so precisely using the assumption of coherent point sources. Typically one rather has to expect an intermediate state between full coherence and random phase. This error results in less pronounced extrema, both minima and maxima, in the measurement compared to the prediction.

Deviations for the low-frequency range at about 500 Hz have the same reason as explained in the frequency response considerations. The second most obvious differences between measurement and computation occur in the crossover range between 2 and 4 kHz, where the first sidelobes are reproduced with an error of about 3 dB. This effect will be investigated further in the following subsection.

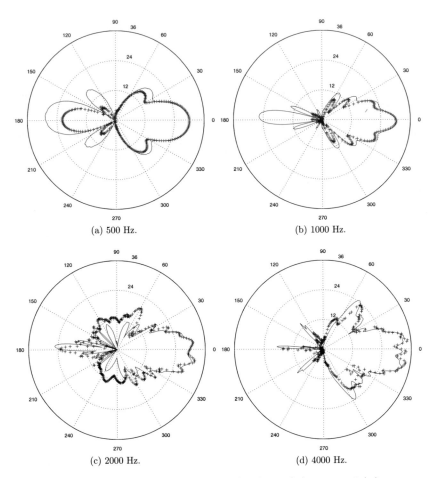

(a) 500 Hz.

(b) 1000 Hz.

(c) 2000 Hz.

(d) 4000 Hz.

Figure 3.13.: Directional response, isolated case (−), measured (+).

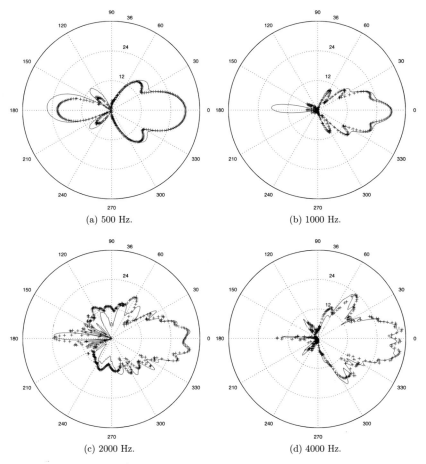

(a) 500 Hz.

(b) 1000 Hz.

(c) 2000 Hz.

(d) 4000 Hz.

Figure 3.14.: Directional response, flanked case (−), measured (+).

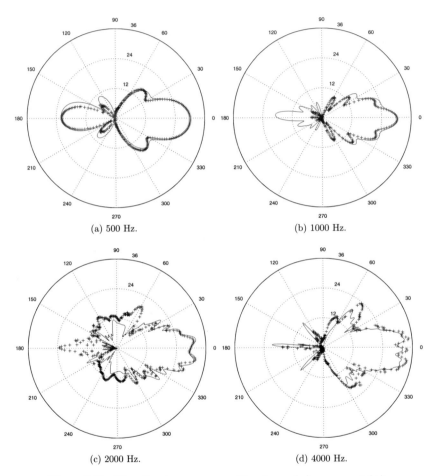

Figure 3.15.: Directional response, positional case (−), measured (+).

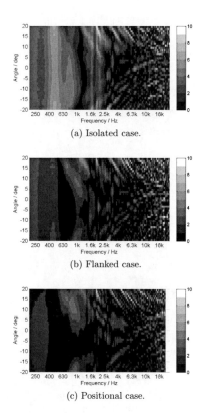

(a) Isolated case.

(b) Flanked case.

(c) Positional case.

Figure 3.16.: Modeled sensitivity relative to measurement. Lighter colors indicate larger deviations.

Elemental Deviation

So far it was assumed that the individual cabinets used in the array have identical axial sensitivities. However, in practice there are some variations between individual samples. Fig. 3.17a shows the deviations of the frequency responses of the 12 different cabinets from the mean of the set. Over the largest part of the frequency range the cabinets fall within a variation of about ±1 dB. Only in the crossover range larger deviations of up to 3 dB occur. Fig. 3.17b displays a similar plot for the on-axis phase responses. The variation is approximately ± 6° with peaks of up to 30° in the crossover range.

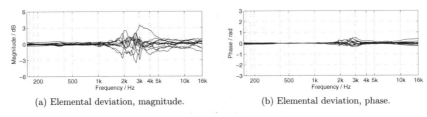

(a) Elemental deviation, magnitude.　　　　　(b) Elemental deviation, phase.

Figure 3.17.: Distribution of elemental responses.

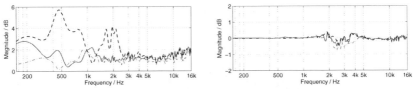

(a) Deviation of isolated (- -), flanked (−), and positional (-.-) case relative to measurement, averaged over ± 20°.

(b) Deviation of isolated (- -), flanked (−), and positional (-.-) case relative to measurement, averaged over ± 20° relative to non-elemental.

Figure 3.18.: Measured and modeled sensitivity including elemental corrections.

For a better understanding of the effect of these errors, the respective point sources were corrected to incorporate the deviations and the whole array was modeled once more. Fig. 3.18a shows the resulting deviations between measurement and prediction. Compared to Fig. 3.12c the behavior seems to be largely unaltered. But a closer look at the crossover range reveals that particularly in this sensitive region the error had decreased notably, on average by about 0.5 dB. This is most obvious in Fig. 3.18b, which quantifies the improvement relative to the measurements when using elemental corrections.

The same effect is shown in Figs. 3.19, 3.20 and 3.21, where the directional data for 2.5 kHz are presented for each model - isolated, flanked, and positional. Compared to the

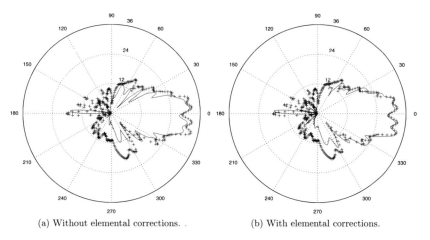

(a) Without elemental corrections. .   (b) With elemental corrections.

Figure 3.19.: Directional response at 2500 Hz, isolated case $(-)$, measured $(+)$.

original polar response comparison the match seems to be much better now, especially the small sidelobes are reproduced clearly.

Of course it needs to be stated that this finding is not of much immediate practical value. In reality it is typically not possible to measure the on-axis responses of all concerned loudspeaker cabinets for prediction purposes. This is especially true when an installation is still being planned and the particular loudspeakers are not even available yet. Nevertheless one gains good insight into the effect of sample-to-sample variations on the prediction accuracy for a full array. Surprisingly it is not as large as one would expect using simple methods of error propagation. This can be explained by the random nature of the variation, as will be shown in Section 3.4. On the other hand, this result quantifies the potential accuracy gains in the simulation that can be obtained by improvements with regard to production tolerance.[4]

It should be added that so far different cabinet samples have not been compared with respect to full directional data. However, it is likely that the axial deviation is dominant compared to the variations in the polar response.

---

[4]In the future, it may be possible to store individual calibration data in the on-board memory of a loudspeaker. Optimization software could then recover this information via the network to improve the accuracy of the result.

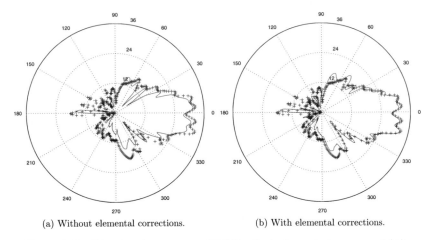

(a) Without elemental corrections. (b) With elemental corrections.

Figure 3.20.: Directional response at 2500 Hz, flanked case (−), measured (+).

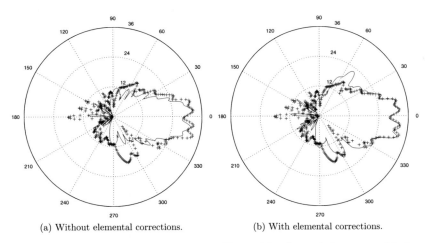

(a) Without elemental corrections. (b) With elemental corrections.

Figure 3.21.: Directional response at 2500 Hz, positional case (−), measured (+).

Figure 3.22.: SEQUENZA line array setup.

3.3.2. Medium-Size Touring Line Array

Several comparisons were made using the SEQUENZA 10 line array. This was specifically selected to show that the prediction methods presented are applicable to larger cabinet sizes, more complex transducer arrangements and beyond the vertical domain [50]. For mechanical reasons a small array of 3 SEQUENZA 10N boxes was used. The front height of a cabinet is about 0.30 m, the overall size of the stacked array is approximately 0.91 m. In this model each cabinet contains three different complex directivity point sources, one each for the LF, LF/MF, and HF transducer. The controller settings for the crossover and EQ were measured separately and also applied in the GLL.

All of the directional measurements were made at a distance of 8 m so that the far-field condition was approximately fulfilled. Full-sphere complex data were acquired. For the single box all three transducer measurements were performed about the same point of rotation, namely, the geometric center of the cabinet. The use of phase data in the prediction automatically accounts for the spatial offset of the transducers relative to the point of rotation (see Section 2.1). The single-box measurements were made without either top or bottom neighbor, which is equivalent to the isolated case, as defined before. The HF transducer was measured with 2° angular resolution, the LF and LF/MF transducers with 5°.

Directional Response

A detailed investigation of the CDPS model in the vertical domain has already been presented, therefore the intention here is mainly to show that the concept works similarly well for any other domain. For this purpose a set of cross-sectional polar plots at the crossover frequency of the system has been chosen. The array presented uses splay angles of 0° between adjacent elements (Fig. 3.22). Other frequencies and configurations were examined as well but did not provide additional insights.

Fig. 3.23a and 3.23b show the horizontal and vertical polar plots at 1250 Hz smoothed to a bandwidth of one-sixth octave for better comparison. The radial scale is 40 dB, and 0° denotes the on-axis direction, 90° the upward direction. Diagonal polar plots are presented in Fig. 3.23c to 3.23f. Four different diagonal planes are shown, namely, 30°/210°, 60°/240°, 120°/300°, and 150°/330° rotated about the system axis in clockwise direction, where the left is 0° when looking out of the box.

The correlation between measured and predicted performances is very good. Within an opening angle of about 50° the deviations are typically within ±2 dB. Because the measurements were made for an isolated cabinet, some of the improvements introduced earlier and affecting off-axis prediction accuracy do not apply here. Therefore the deviations in the polar responses increase to about ±3 dB for larger angles. Overall one can state that the simulation matches the measurement very well, and this includes planes other than horizontal and vertical.

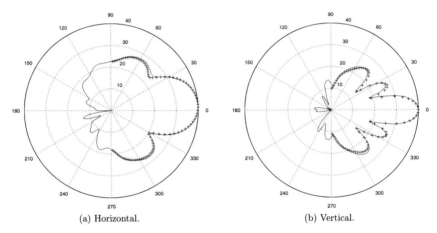

(a) Horizontal.   (b) Vertical.

Figure 3.23.: Directional response at 1250 Hz; measured (+) and modeled (−).

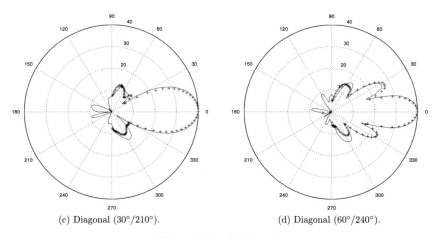

(c) Diagonal (30°/210°). (d) Diagonal (60°/240°).

Figure 3.23.: *Continued.*

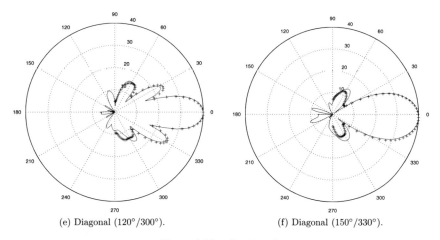

(e) Diagonal (120°/300°). (f) Diagonal (150°/330°).

Figure 3.23.: *Continued.*

3.4. Sample Variation

The previous sections have been concerned with the accuracy of line array modeling and corresponding data resolution and measurement requirements. Based on that, in this section the influence of variations among supposedly identical components of an array on the overall performance of the system [119] will be discussed.

This topic is of high importance since on the one hand it allows investigating the spread of acoustic performance quantities over a range of mechanically identical line arrays. On the other hand it allows estimating the error due to assuming that individual array elements contribute equivalently to the radiated sound field of the whole array. Similarly important, it gives some insight into the errors made when modeling such an array based on the assumption of identical components, especially with respect to their on-axis response and directional pattern.

At first glance, one may think that detailed loudspeaker performance data, such as high-resolution magnitude data or phase data, are generally not useful because the production variance of a loudspeaker series as well as measurement errors cause these details to be irrelevant. In this section it is shown that the effect of sample variation is not necessarily significant for prediction accuracy. On the contrary, the actual modeling error decreases with the number of partial sound sources in the system. High-resolution loudspeaker data reduce the error in the final result of the simulation.

The aforementioned conclusions are derived from three alternative approaches:

- A rigorous mathematical analysis is performed based on statistical distribution functions for the individual elements and on the central limit theorem.

- Numerical simulations of typical, idealized cases are analyzed utilizing scattering properties with different sample distributions and array sizes.

- It is demonstrated that the theoretical results agree with measurement data and simulation results from several real-world loudspeaker systems.

3.4.1. Mathematical Background and Simulation Results

Magnitude Variation

The analysis begins with the array of point sources given by eq. (1.4) based on point sources j as defined by eq. (1.7). For initial investigations the focus will primarily be on a linear arrangement of sources, such as eq. (3.20), and on receiver angles roughly perpendicular to the extension of the line source, that is, on-axis so that $A_j \approx A_j(\omega)$ only depends on frequency:

$$p_{sum}(\vec{r}, \omega) = \sum_{j=1}^{N} \frac{A_j(\omega)}{|\vec{r} - \vec{r}_j|} e^{-ik|\vec{r} - \vec{r}_j|}. \tag{3.30}$$

It is assumed that in the simplest case these point sources are located close to the origin whereas the receiver is in the far field, $r = |\vec{r}| \gg |\vec{r}_j|$. At this point, phase differences between individual sources caused by receiver locations slightly off the horizontal plane ($\theta \approx 0°$ in Fig. 3.1) will be small enough at the frequencies of interest to be ignored: $k|\vec{r} - \vec{r_j}| \approx \frac{2\pi}{\lambda}(r - x_j \sin\theta_j) \approx \frac{2\pi}{\lambda}r$. As a result, one may rewrite the relationship (3.30) in the following form:

$$p_{sum}(\vec{r}, \omega) = \frac{1}{r}e^{-i\frac{2\pi}{\lambda}r}\sum_{j=1}^{N} A_j(\omega). \qquad (3.31)$$

This may be further simplified by assuming that the point sources show identical phase behavior[5], $A_j \equiv |A_j|$, and that a common relative phase can be removed without loss of generality, $p_{sum} \equiv |p_{sum}|$,

$$p_{sum}(\vec{r}, \omega) = \frac{1}{r}\sum_{j=1}^{N} A_j(\omega). \qquad (3.32)$$

One may now look at the effect of a statistical variation among the sources in this simplified tutorial example. For this purpose it can be assumed that all point sources have approximately the same radiation function A_j and that their spread is governed by a statistical distribution function. As will be shown later, there are good reasons to assume that in practice the properties of most loudspeakers or transducers of the same make can be represented by a normal (Gaussian) distribution:

$$W(A) = \frac{1}{\sqrt{2\pi}\delta A}\exp\frac{-(A - \langle A\rangle)^2}{2(\delta A)^2}, \qquad (3.33)$$

where A represents a possible value for the magnitude A_j, $\langle A \rangle$ is the mean magnitude and δA is the standard deviation. The function $W(A)$ then defines the probability density for a particular value of A. That means that $W(A)dA$ is the probability for a given source j to have a pressure amplitude A_j located in the interval $[A, A + dA]$.

The central limit theorem [120] states that the sum over a set of N random processes i, each defined by

$$Z_i \sim W_i(\mu_i, \sigma_i^2), \qquad (3.34)$$

with W_i being the particular distribution function governed by mean μ_i and standard deviation σ_i, yields a random process with a Gaussian distribution G of the form:

$$Z \sim G\left(\sum_i \mu_i, \sum_i \sigma_i^2\right) \qquad (3.35)$$

One can consider eq. (3.32) as such a sum of individual, independent random contributions

$$p_i = \frac{1}{r}A_i. \qquad (3.36)$$

Each of them has a mean value of

$$\mu_i = \langle p_i\rangle = \frac{1}{r}\langle A_i\rangle = \frac{1}{r}\langle A\rangle, \qquad (3.37)$$

---

[5]Phase variations will be discussed further below.

and a variance of

$$\sigma_i^2 = (\delta p_i)^2 = \langle p_i^2 \rangle - \langle p_i \rangle^2 = \frac{1}{r^2}(\delta A_i)^2 = \frac{1}{r^2}(\delta A)^2. \tag{3.38}$$

Accordingly one finds for the probability function of the sum in eq. (3.32) the relation

$$Z \sim G\left(\frac{N}{r}\langle A \rangle, \frac{N}{r^2}(\delta A)^2\right), \tag{3.39}$$

with mean

$$\langle p_{sum} \rangle = \mu_{sum} = \frac{N}{r}\langle A \rangle, \tag{3.40}$$

and standard deviation

$$\delta p_{sum} = \sigma_{sum} = \frac{\sqrt{N}}{r}\delta A. \tag{3.41}$$

This yields the relative error of the array:

$$\frac{\delta p_{sum}}{\langle p_{sum} \rangle} = \frac{1}{\sqrt{N}}\frac{\delta A}{\langle A \rangle}. \tag{3.42}$$

This means that the more sources are combined together the lower the error of the resulting pressure field will be, considered relative to the error of the individual source.

In the following considerations, the relative source error in dB is utilized as a typical parameter. It is here defined as

$$L\left(\frac{\delta A}{\langle A \rangle}\right) = 20\log\left(\frac{\delta A}{\langle A \rangle} + 1\right), \tag{3.43}$$

where $\frac{\delta A}{\langle A \rangle}$ is the (linear) relative error. For small errors one can assume that

$$L\left(\frac{\delta A}{\langle A \rangle}\right) \approx \frac{20}{\ln 10}\frac{\delta A}{\langle A \rangle}. \tag{3.44}$$

The relative array error in dB is defined in the same way, as $L\left(\frac{\delta p_{sum}}{\langle p_{sum} \rangle}\right)$. Note that in practice, absolute dB values are often accompanied by an error estimate given in $\pm$dB. This corresponds to a multiplicative error in the linear domain and can be related directly to the relative error only when the value is small, approximately ≤ 1 dB.

In Fig. 3.24 this analytical result is illustrated and compared with simulation results that are based on a large set of random line array realizations. Fig. 3.24a depicts the dependency of the relative error of the array of combined sources on the number of sources. A typical value of $L\left(\frac{\delta A}{\langle A \rangle}\right) = 1.0$ dB was used for the error of the single source. Two different receiver locations were used, on-axis and at $5°$ off-axis. The angular parameter of $5°$ was only accounted for in the simulation run, therefore the corresponding theoretical result was offset accordingly but without changing the curve's

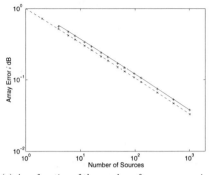 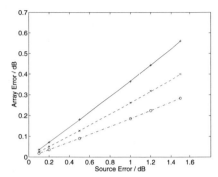

(a) As a function of the number of sources, on-axis simulated ($\times$) and calculated (- -), 5° off-axis simulated ($+$) and calculated ($-$), for a source error of 1 dB.

(b) As a function of the source error, 8 sources simulated ($+$) and calculated ($-$), 16 sources simulated ($\times$) and calculated (- -), 32 sources simulated ($\circ$) and calculated (-.-).

Figure 3.24.: Relative error of array of sources, displayed in dB according to eq. (3.43) here and in the following figures.

slope. For the 5°-curve, the error is expectedly slightly higher because the sources do not combine ideally anymore due to relative phase shifts caused by travel time differences. For both angles, the error in the simulation decreases with an increase of the number of elements in the array and scales with $1/\sqrt{N}$ exactly as predicted. Fig. 3.24b shows a comparison between simulation results and theory for three different array sizes N. The error increases linearly with the error of the single source, both in simulation and theory. A larger source count results in a smaller error.

Level-Shaded Array

In practice line arrays consist of multiple boxes and transducers. Even within the rated coverage angle that encloses the on-axis direction of the transducer or cabinet, the radiation of the individual source will be slightly angle-dependent. Also, different filter settings may be applied to the sources, such as used in amplitude or frequency shading applications[6]. For a given receiver location these effects can be described by introducing a weighting factor h_j for each source j in equation (3.32):

$$p_{sum}(\vec{r}, \omega) = \frac{1}{r} \sum_{j=1}^{N} h_j A_j(\omega). \tag{3.45}$$

Like in the previous section it is now assumed that the radiation function A_j of each source j is a random quantity that is described by a Gaussian distribution as defined by

---

[6]Shading filters are often used to shape the directional pattern of a loudspeaker array as a function of distance or frequency, respectively.

eq. (3.33), with mean $\langle A_j \rangle = \langle A \rangle$ and standard deviation $\delta A_j = \delta A$. Based on that one may calculate the statistical properties of the individual pressure functions $p_j = \frac{1}{r} h_l A_j$ and then, by applying the central limit theorem again, of the sum pressure function $p_{sum} = \sum_{j=1}^{N} p_j$. This yields for the mean of the sum

$$\langle p_{sum} \rangle = \frac{1}{r} \sum_{j=1}^{N} h_j \langle A \rangle, \tag{3.46}$$

and for the standard deviation

$$\delta p_{sum} = \frac{1}{r} \sqrt{\sum_{j=1}^{N} h_j^2 \delta A}. \tag{3.47}$$

This allows deriving the upper boundary for the resulting relative error because the asymptotic limit is located where only a single source j dominates the pressure sum with $\sum_j h_j^2 \approx h_j^2$:

$$\frac{1}{\sqrt{N}} \frac{\delta A}{\langle A \rangle} \leq \frac{\delta p_{sum}}{\langle p_{sum} \rangle} \leq \frac{\delta A}{\langle A \rangle}. \tag{3.48}$$

Evidently, the overall relative error can be maximally as large as the error of a single source.

Figure 3.25 shows a comparison between simulation results and theory for 3 different array sizes N as a function of the filter slope. Symmetrical, linear amplitude shading was applied relative to the center of the array, which means that outer elements in the array are more attenuated than center elements. The error of the single source was assumed to be $L(\frac{\delta A}{\langle A \rangle}) = 1.0$ dB.

For small amplitude shading the error decreases with an increase in the number of elements in the array and scales with $1/\sqrt{N}$; for large amplitude shading the error approaches the error of the single, dominant source. Similarly, for a larger number of sources, that is, a higher density, more sources will contribute to the resulting sum than for a smaller number when looking at the same value for the filter slope. This in turn leads to a more gentle increase of the array error with greater filter slope.

Phase Error

Still looking at the effect of sample variation on the radiation in the main direction one may assume that there is a variation of the phase as well. While this can be treated with respect to real and imaginary part like two independent forms of equations (3.32) and (3.39), it is primarily interesting to evaluate the error of the resulting magnitude data which is a function of both error in elemental magnitude and in elemental phase.

In order to estimate the influence of the phase error $\delta\phi$, eq. (3.31) is used again. One may assume a vanishing magnitude variation δA and that any constant phase offset can be neglected without loss of generality:

$$p_{sum}(\vec{r}, \omega) = \frac{A_0}{r} \sum_{j=1}^{N} e^{i\phi_j}, \tag{3.49}$$

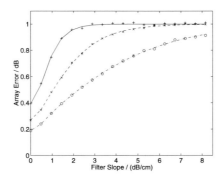

Figure 3.25.: Relative error of array of sources as a function of symmetrical amplitude shading and of a source error of 1 dB, 7 sources simulated (+) and calculated (−), 15 sources simulated (×) and calculated (- -), 31 sources simulated (∘) and calculated (-.-).

where A_0 is the constant magnitude part of the radiation amplitude A_j of each source and $\phi_j(\omega) = \arg A_j$ is the phase part of the complex amplitude. One may choose $\langle \phi_j \rangle = \langle \phi \rangle = 0$ and assume that the variation of phase is small, $\phi_j \ll 1$. A Taylor approximation keeping only terms of lowest order yields

$$p_{sum}(\vec{r}, \omega) \approx \frac{A_0}{r} \sum_{j=1}^{N} \left(1 - \frac{\phi_j^2}{2} \right), \tag{3.50}$$

if one also assumes N to be large enough so that $\sum_j \phi_j \ll \sum_j |\phi_j|$. In consequence the mean of each summand is

$$\langle p_j \rangle = \frac{A_0}{r} \left(1 - \frac{(\delta\phi)^2}{2} \right), \tag{3.51}$$

and its variance

$$(\delta p_j)^2 = \frac{A_0^2}{2r^2} (\delta\phi)^4. \tag{3.52}$$

Accordingly the sum error yields for first-order terms:

$$\frac{\delta p_{sum}}{\langle p_{sum} \rangle} = \frac{(\delta\phi)^2}{\sqrt{2N}}. \tag{3.53}$$

Figure 3.26a shows a comparison between simulation results and theory as a function of the number of sources N. Similarly, Fig. 3.26b compares the simulated and the theoretical array error as a function of the phase error $\delta\phi$. The results deviate only slightly and only for small amounts of sources and large source errors. This is where the accuracy of the theoretical approximation is limited.

It should be emphasized that the effect of phase variations on the resulting sum scales with the square compared to the magnitude error which is linearly proportional. Thus

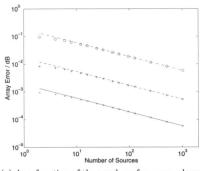

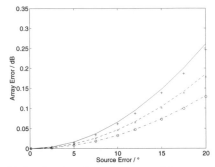

(a) As a function of the number of sources, phase error of 10° simulated (○) and calculated (-.-), phase error of 3° simulated (+) and calculated (- -), phase error of 1° simulated (×) and calculated (−).

(b) As a function of the phase error of the source, 8 sources simulated (+) and calculated (−), 16 sources simulated (×) and calculated (- -), 32 sources simulated (○) and calculated (-.-).

Figure 3.26.: Relative error of array of sources.

for small phase errors the effect will usually be insignificant compared to the magnitude error. For example, for a typical phase error of 10° and 4 sources the resulting relative array error is already smaller than 0.1 dB.

Array Directivity

Finally the effect of sample variation on the directivity of a line array should be investigated. The main concern here is the mean value and the standard deviation of the directional function of the statistical array compared to the ideal array, under the condition of a varying source magnitude.

Figure 3.27 shows the computational results for the vertical directivity of a line array of 16 omni-directional sources and of a length of 1 m, simulated in the far field and at a frequency of 2 kHz. The normalization for the level is chosen such that the on-axis level is 40 dB. For display purposes, the relative array error was offset by 30 dB, higher values indicate a non-zero error in dB. Mean and standard deviation were derived based on the stochastic simulation of a set of 10 000 arrays.

Fig. 3.27a demonstrates that a relative error of 1 dB in the magnitude of each source has very small effect on the directional behavior of a line array. The average array is very close to the ideal array. The only significant exception concerns the reproduction of the minima. This can be explained by the fact that the variation in the magnitude of the individual source will not allow the sources in the array to combine perfectly at a given angle and cancel each other out exactly.

The relative error is small for on-axis directions and only increases in the proximity of the nulls. However, the latter is of no practical significance since for these angles

114

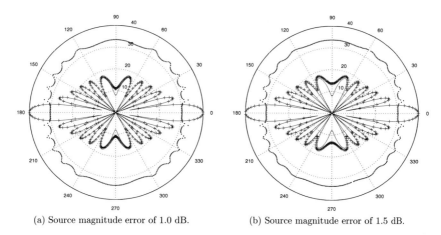

(a) Source magnitude error of 1.0 dB.   (b) Source magnitude error of 1.5 dB.

Figure 3.27.: Vertical directivity and error of line array at 2 kHz, ideal array $(-)$, average array with source magnitude variation $(+)$, relative error $(\cdots)$. (The latter is offset by 30 dB.)

the level is very low, anyway. For off-axis angles the relative error increases up to 3 dB, which can be understood in the frame of the analytical results presented above. Here phase differences between elements become significant and affect the combination of sources. An in-depth mathematical analysis of this result is complicated and requires future work.

It is illustrated in Fig. 3.27b that the qualitative behavior does not change for a relative error of 1.5 dB. Only the general deviation and array error increase slightly. It should be remarked that a magnitude scatter of 1.5 dB among sources is overestimated compared to the small variation within high-quality series in reality.

Finally, Fig. 3.28 displays the array error averaged over the vertical angle, as a function of frequency. At low frequencies the array error is small. In this range the radiation pattern of the array is nearly omnidirectional and the average error is roughly equal to the on-axis error, in this case 0.25 dB for $N = 16$ elements. Toward higher frequencies the error is increasing, primarily due to differences at angles where the directional function of the ideal array is very small and that of the statistical array is slightly smoothed by the randomized source amplitudes. For this reason the average error will depend on the number and the relative width of these minima. Compared to the far field, the mean error at distances close to the array is slightly smaller due to less pronounced cancelations in the ideal directional response.

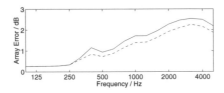

Figure 3.28.: Relative error of array of sources as a function of frequency, for a source error of 1 dB, 16 sources, at a distance of 2 m (- -) and in the far field (—).

Practical Remarks

Simulation results were computed using a standard Monte Carlo approach [121]. To derive statistical data for a set of arrays, frequency response data for each source in each array was generated using the Box-Muller method assuming a Gaussian (normal) distribution of data [122]. All stochastic simulations were performed using a line array of 1 m length, at 2 kHz, with the receiver located in the far field. A set of 10 000 arrays was simulated to obtain the statistical properties.

As a side note it should be mentioned that this study is primarily concerned with the effect of variation among the frequency response or sensitivity of individual sources. The variation of the directional response of the sources was not considered, because experience indicates that these effects are of secondary order (see also Section 3.3).

3.4.2. Comparison with Measurement

To evaluate the match between theory and practice, measurement results will now be compared with results predicted according to the above theoretical derivations. Data will be used from a compact modular line array (Omniline by Martin Audio [116], as in Fig. 3.10), from a column loudspeaker (Iconyx IC-8 by Renkus-Heinz [123], similar to Fig. 2.40c), and from a medium-size touring line array (Electro-Voice XLC [124], comparable to the cabinets in Fig. 3.22).

For each of those, a large set of on-axis frequency responses for the individual sources, that is, cabinets or transducers, respectively, was measured. The statistical distribution of magnitude and phase data of the sources was then analyzed. It is found that the distribution functions are approximately Gaussian. Additionally, the phase error is found to be small enough to be neglected when comparing its effect with that of the magnitude error.

For the compact array and the column loudspeaker a data set for the assembled arrays was measured as well. Based on these measurement data relative errors for the arrays are calculated and compared with the theoretically expected results derived from the relative error for the sources.

Compact Line Array

Figure 3.29a shows the frequency responses of 423 Omniline cabinets, measured on-axis and normalized to the mean of set. The corresponding relative error of these magnitude data is presented in Fig. 3.29b. It should be noted that the increased relative error in the frequency range of 2-4 kHz is primarily due to the crossover where the variation of loudspeaker samples is naturally larger (see also Section 3.3).

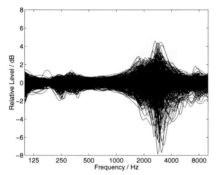 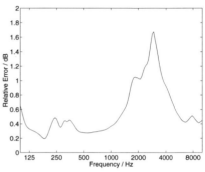

(a) Frequency response measurements of single cabinets, normalized to mean of set.

(b) Relative error of cabinet magnitude data.

Figure 3.29.: Statistical properties of Omniline cabinet samples.

The data for the third and fourth moment [125] indicate that the distribution is close to Gaussian. As an example, Fig. 3.29c shows the relative distribution function for a frequency of 1 kHz. Skewness and excess are depicted in Figs. A.2a and A.2b in Appendix A.3.

Similarly, the absolute error of the phase data is displayed in Fig. 3.29d, higher moments in Figs. A.2c and A.2d. Also the phase data obeys a Gaussian distribution function. The absolute phase error rarely exceeds 10°. For this reason one can expect that the phase variation will not influence the end result very much, compared to the magnitude variation.

Now an array of such cabinets should be looked at. Fig. 3.30 shows the distribution and the statistical properties of a set of four Omniline arrays, each consisting of $N = 8$ cabinets in a straight configuration. The relative error for the low- and mid-frequencies has reduced in average from 0.4 dB to below 0.2 dB. This is a little bit less than the expected factor of $1/\sqrt{N}$, but it is still satisfyingly close (Fig. 3.30b). In the crossover range and above the deviation is larger; the relative array error is roughly equal to the source error. Increased influence of source directionality as well as of propagation phase on the pressure sum at the receiver may contribute to that and are not accounted for by the model eq. (3.42).

Summing up one may state that although a number of four items may not be enough to establish a statistically reliable data set, the match is still remarkably good.

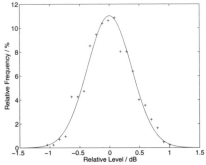

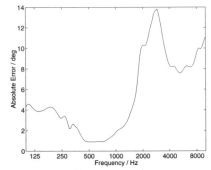

(c) Relative frequency of levels at 1 kHz, measured cabinet data (+) and exact Gaussian distribution (−).

(d) Absolute error of cabinet phase data.

Figure 3.29.: *Continued.*

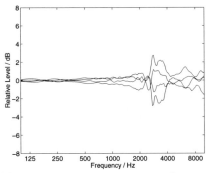

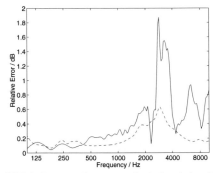

(a) Frequency response measurements of arrays, normalized to mean of set.

(b) Relative error of array magnitude data (−) and theoretically expected result (- -), see text.

Figure 3.30.: Statistical properties of Omniline array samples.

3. Modeling Large Sound Sources

Loudspeaker Column

The second example is not based on a mechanically configurable line array but on a column loudspeaker with eight identical transducers. Figure 3.31a shows the on-axis frequency responses of 72 Iconyx (IC-8) transducers, normalized to the mean of the set. Fig. 3.31b displays the relative error of the same. Figs. A.4a and A.4b in Appendix A.3 depict the third and fourth moment, respectively, and illustrate that the assumption of a Gaussian distribution is reasonable. As an example, Figs. 3.31c and 3.31d show the distribution function for 500 Hz and 1 kHz, respectively.

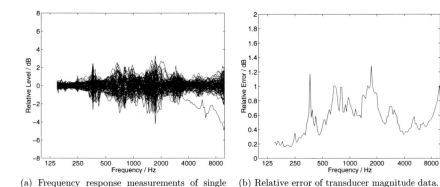

(a) Frequency response measurements of single transducers, normalized to mean of set.

(b) Relative error of transducer magnitude data.

Figure 3.31.: Statistical properties of IC-8 transducer samples.

The frequency dependence of the phase variation for the same set of transducers is depicted in Fig. 3.31e. It is generally below 10°. Also here, the phase error is negligible compared to the magnitude error. The other statistical characteristics of the phase data are presented in Figs. A.4c and A.4d.

The 72 transducers stem from nine different IC-8 column loudspeakers, each comprising $N = 8$ transducers. Fig. 3.32a shows the frequency responses and Fig. 3.32b the standard deviation for the entire loudspeakers. It can be recognized that the relative error in the mid-frequency range decreases on average from about 0.7 dB down to 0.25 dB. This corresponds roughly to the factor of $1/\sqrt{N}$ by which the relative error should reduce theoretically.

The distribution for the array shows smaller skewness (Fig. A.5a) and excess (Fig. A.5b) than the transducer data, as to be expected by the central limit theorem. The more random processes are summed, the closer to a Gaussian distribution the result will be.

In the high-frequency range the relative error does not decrease as much. This can probably be attributed to the fact that the measurements were taken in the transition zone to the far field. Here propagation phase differences between individual transducers

119

become more significant. Also, due to their more pronounced directionality at high frequencies the transducers do not sum up equally at the receiver location. For both reasons the simple theoretical model expressed by eq. (3.42) is not very accurate in this case.

Towards lower frequencies a different effect occurs. In this frequency range the data for the individual transducers are already close to the noise floor of the measurement and to the reproduction accuracy, which are of the order of 0.15 dB. The uncertainty of the array measurements is even larger and thus cannot show the full reduction of relative error. Depending on the nature of the noise floor, additional averaging of raw measurement data may be helpful.

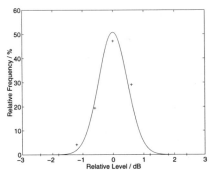
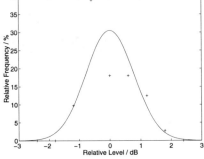

(c) Relative frequency of levels at 500 Hz, measured transducer data (+) and exact Gaussian distribution (−).

(d) Relative frequency of levels at 1 kHz, measured transducer data (+) and exact Gaussian distribution (−).

Figure 3.31.: *Continued.*

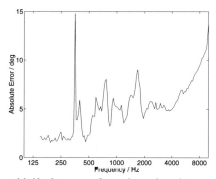

(e) Absolute error of transducer phase data.

Figure 3.31.: *Continued.*

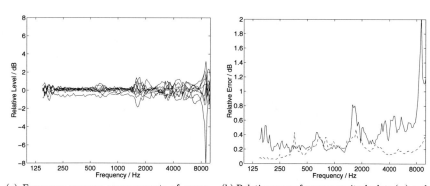

(a) Frequency response measurements of arrays, normalized to mean of set.

(b) Relative error of array magnitude data (−) and theoretically expected result (- -), see text.

Figure 3.32.: Statistical properties of Iconyx IC-8 array samples.

121

Medium-Size Touring Line Array

The XLC line array from Electro-Voice is a high-performance three-way system of medium cabinet size. Figures 3.33, 3.34, and 3.35 show the distribution of the on-axis frequency responses of 34 measured cabinets for the LF, MF, and HF pass-band, respectively:

- Fig. 3.33a and 3.33b indicate an approximate relative error of 0.25 dB for the LF section over its bandwidth of normal operation from 20 Hz to 200 Hz.

- The operational bandwidth of the MF unit is approximately from 200 Hz to 1200 Hz. The typical relative error over this frequency range is about 0.3 dB as shown in Figs. 3.34a and 3.34b.

- The level distribution of the HF pass-band is displayed in Figs. 3.35a and 3.35b. The average relative error over its bandwidth from 1200 Hz to 10 kHz is about 0.35 dB.

More detailed statistical analysis reveals that the distribution functions of the three magnitude data sets can be considered Gaussian. The normalized skewness and excess for all three pass-bands rarely exceed a value of 1. The absolute error of the phase data is generally below 10°, for the HF unit even below 5°. Skewness and excess the phase data do not exceed a value of 2, typically.

Even though XLC array measurements are not available these statistical results confirm the assumption previously made that the magnitude error is normally well below 1 dB and the phase error below 10°. It can be safely assumed that both distributions are Gaussian.

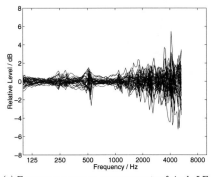

(a) Frequency response measurements of single LF pass-bands, normalized to mean of set.

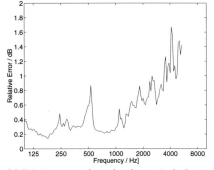

(b) Relative error of pass-band magnitude data.

Figure 3.33.: Statistical properties of XLC LF pass-band samples.

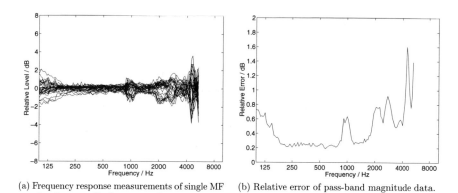

(a) Frequency response measurements of single MF pass-bands, normalized to mean of set.

(b) Relative error of pass-band magnitude data.

Figure 3.34.: Statistical properties of XLC MF pass-band samples.

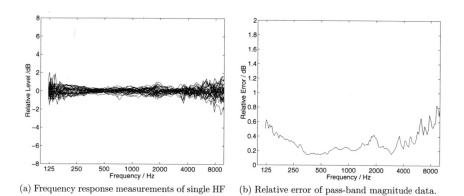

(a) Frequency response measurements of single HF pass-bands, normalized to mean of set.

(b) Relative error of pass-band magnitude data.

Figure 3.35.: Statistical properties of XLC HF pass-band samples.

Remarks

In the above analysis the focus was primarily on the effect of the magnitude error due to sample variation in the array. It was shown that relative to that the phase error is largely negligible for the given sample data. One may perform investigations similar to that of the magnitude error for the real and imaginary part as well. Within the same error limits they also obey a Gaussian distribution. However, this is not a necessary requirement for the application of the central limit theorem. The fact that the data set of the single sources already satisfies a Gaussian distribution indicates that there are multiple underlying random processes that combine together in this distribution according to the central limit theorem.

The large but still limited amount of data that were available for this study supports the theoretical expectations. The relative error of the array is generally lower than the relative error of the source magnitude, and obeys a $1/\sqrt{N}$-law when the sources are approximately of equal strength and phase at the receiver. Currently on-going investigations related to advanced digital beam-steering relying heavily on FIR filters also reveal no substantial negative effects of sample variation on the overall array performance [126].

3.5. Summary and Open Problems

In this chapter it was shown that the CDPS model can quite naturally be extended to large sound sources such as column loudspeakers or line arrays. Modeling results based on the CDPS decomposition are highly accurate if resolution requirements are met and the acoustic support of neighbor cabinets in a line array is accounted for. Sample variation among cabinets or transducers of the same type can normally be neglected when data of a representative or average unit is used. In fact, statistical error compensation effects lead to a smaller scatter of mechanically identical arrays than of their components.

On this basis the CDPS model allows replacing other modeling philosophies that are for example based on the reproduction of the wave front by elementary sources or by a tesselated, virtual surface. However, there are a number of open issues that should be pointed out:

- It is an important fact that CDPS decomposition can only be applied to line sources that can be subdivided and measured as partial sources. The CDPS model cannot be easily employed for large continuous sources, such as ribbon loudspeakers with a length of 2 m or large horns. In these cases, it is physically impossible to decompose the original source into several components that can be measured in their far field. This lack can be circumvented to some extent by e.g. making scale model measurements or relying on a mathematical model in order to generate the complex directional data for the CDPS approach.

- Another important aspect is that the CDPS model assumes fully coherent sources in a time-invariant system. However, in practice this assumption has limits due to fluctuating environmental conditions, such as air movements. As will be discussed next, in Chapter 4, coherence among sources can be affected by local, temporary changes of the propagation medium air. Coherence is reduced at larger receive distances and at higher frequencies.

- It is also worth mentioning that the CDPS model being based on the superposition of point sources cannot account for boundary effects, such as shadowing, coupling or diffraction by neighbor cabinets in the array (as discussed e.g. in [127]), unless these effects are already included in the measurements.

Beyond that attention should also be given to a number of aspects that are important from the practical perspective:

- The accurate measurement of directional magnitude and phase data in the appropriate resolution is crucial for the process. Measurement data must be reproducible within a defined error margin, with respect to both measurements of the same cabinet and measurements of different cabinets of the same type. In this regard, sensitivity data and directional data used for the CDPS model should be complemented by information that determines the uncertainty of the data quantitatively.

- As it has been explained earlier, the computational performance of the CDPS model is significantly higher than any approach that is based on some sort of elementary sources which are subject to spatial resolution requirements that scale with the wavelength. Nonetheless the CDPS model is also measurably slower than using a single directional data set measured in the far field of the array. Practice has shown that the computational load is dominated by the transforms from magnitude and phase data to real and imaginary parts and back. Both transforms are inevitable because the directional data is stored and interpolated as magnitude and phase data. In contrast, the complex summation of sources is performed using real and imaginary parts. Finally, most end results are again based on magnitude data.

- In practice required computer memory is rarely an issue. In terms of a computer model, a typical loudspeaker GLL file seldom exceeds the size of about 1 MB of data. However, line array GLL files that contain multiple cabinet definitions, especially when they use directional data with angular resolutions higher than 5°, may be as large as 10 MB and more.

- In Section 2.3.2 the necessity for creating source groups in order to perform geometrical visibility tests (Figs. 2.21a, 2.21b) has been discussed. For large sound sources this procedure has also proven to be a good compromise between performance on the one hand and accuracy on the other hand. Particularly in the case of large line arrays that consist of dozens of sources visibility tests for every source individually would lead to impractically long computation times.

Other than that, all solutions that have been developed earlier for small sources apply to large sources as well. Crossover, equalization filters or beam steering filters can be applied as already shown in Section 2.4.4. Mechanical configurability is also given, as demonstrated with the Omniline line array in Section 3.3.1.

4. Coherence of Radiated Sound

4.1. Introduction

In the previous chapter modeling of loudspeaker line arrays has been discussed in great detail. In reality, most sound systems consist of a number of line arrays or other complex loudspeaker systems. These systems are generally difficult to set up and to model. This applies particularly to the proper alignment of all loudspeaker signals in time because the related even coverage of the audience zones and smooth system response are important performance criteria.

However, for the design of such systems there is still no consistent approach concerning the expected superposition of sound signals at the receiver. It is usually assumed that sound sources which are located close to each other combine in a coherent manner, that is, their sound pressure amplitudes are added by taking into account phase information. In contrast, sound sources located far apart from each other are assumed to show a random-phase relationship and therefore sum energetically. In reality, these assumptions only represent the extreme cases but they are not very accurate at all.

In order to predict the coherence of combinations of sound sources, environmental conditions, such as temperature fluctuations or variations of air flow as well as their spatial correlation, have to be accounted for. These parameters determine the level of coherence among sources and thus influence the performance of the sound system as well as the perceived quality, spectral shape, and the temporal stability of the acoustic signal delivered to the receiver.

Except for the above-mentioned differentiation between complex sum and power sum of pressure amplitudes, fluctuating environmental conditions are at large not accounted for in electro-acoustic and room acoustic modeling. Publications relating to the theoretical background or to systematic measurements are scarce. With respect to static environmental conditions, it is common to consider the influence of air temperature, humidity and pressure on the speed of sound and on the attenuation of propagating sound waves, e.g. as defined in ISO 9613 [58].

Significantly more work was and is being done in the field of modeling outdoor sound propagation, especially with respect to noise sources, such as railways or highways, e.g. [128], [61], [62], [129]. Outdoor models must usually account for wind and temperature gradients in the transmission medium air. To some extent these models also consider fluctuations of wind and temperature, such as turbulence. But in contrast to sound reinforcement problems, these discussions are normally limited to a single source or to a source and the ground reflection. Outdoor modeling is typically concerned with large propagation distances and a large range of wind speeds, since the primary interest is in estimating for a given scenario if noise sources are audible and, if so, at which level. However, in electro-acoustics the quality of the received sound plays the dominant role and for this reason only small fluctuations of environmental conditions and comparably small propagation distances will be important. Beyond that the sound will be so much distorted that typical quality criteria become meaningless. For example, strong winds of varying direction and speed affect an outdoor musical performance so much that e.g. SPL coverage and spectral balance over the audience zones are rather unimportant because of their irregular and unpredictable behavior.

Mathematical frameworks for the analytical treatment of sound scattering by inhomogeneities and anisotropies in a fluid medium were developed by numerous authors, such as [130], [3]. Unfortunately, because of their very general approach those provide only limited insight for practical applications.

The gap between these models and practical simulation of outdoor noise propagation is bridged by a number of works in the field of atmospheric acoustics ([131], [132], [133], [134], [65], [135], [136], [137]), of which Ostashev [65] represents a comprehensive summary. His detailed work is concerned with the accurate theoretical treatment of sound propagation in a moving random medium. It is based on solving the perturbed wave equation and provides a good picture of the overall problem of the effect of turbulence on the transmission of sound waves. Nonetheless, due its wave-based approach it lacks a number of aspects that are typical for considerations in the high-frequency limit (Eikonal equation [1], [4], [130]) which is commonly used in the field of electro-acoustic and room acoustic modeling. Beyond that, Ostashev's advanced physical and mathematical presentation can be considered challenging for most practitioners.

In contrast, with directly applicable formulas the work presented here focuses on a selected part of the problem and it includes measurement results, as well. In particular, it is of interest to investigate the combined performance of multiple sources in a slightly inhomogeneous medium. The results obtained here can be considered complementary to the ones presented by Ostashev [65] and provide similar answers in regions of their mutual overlap.

Overview

In this chapter a specific mathematical and physical framework will be developed for the practical inclusion of environmental parameters in the propagation model of the radiated sound field. Based on that, the level of coherence of the elements of the sound system can be quantified using environmental parameters as an input. The theoretical treatment of the problem consists of the following steps:

1. The travel time[1] of the sound wave in a slightly inhomogeneous, anisotropic medium according to eq. (1.6) is used as the starting point.

---

[1]Here and further on the terms *travel time, arrival time,* and *propagation delay* will be used synonymously for the *propagation time* of the signal from the source to the receiver.

2. The spatial dependence of fluctuations of the environmental parameters wind speed and temperature is analyzed by considering the underlying physical processes and expressed as an average over a statistical ensemble [138], [139], [140].

3. Using stochastic theory the standard deviation of the arrival time can then be computed as a function of the source-receiver distance, the amplitude of the fluctuations, as well as the spatial correlation length of the fluctuations.

4. This result can be extended to the standard deviation of the differential arrival time of two sources by including correlation effects between the two propagation paths.

5. The obtained statistical distribution function of the arrival time is used to determine the average magnitude of the interference term in the complex sum of sound pressures and thus the resulting sound pressure level.

This purely analytical treatment provides the important results of this chapter, namely a set of comparably simple mathematical expressions that determine the combined sound pressure level of two or more sources as a function of the amplitude of the fluctuations of the environmental parameters and their spatial correlations. Rather than being a Boolean on/off-function, the newly derived formula describes a continuous transition from power summation on the one end to the superposition of complex pressure amplitudes on the other end, in dependence on the distance between source and receiver, on the geometry of the setup, as well as on the frequency.

After that, experimental data are presented. For a number of different measurement setups ultrasonic anemometers were used to determine statistical properties of wind speed, wind direction, and air temperature. At the same time acoustic impulse response measurements were made to permit later comparison of predicted sound propagation characteristics with measurement results. It is shown that the theoretical results match well with the experimental data under the given limitations. Finally it is illustrated that the presented model also shows satisfying agreement with the solutions provided by Ostashev [65] in the parameter region of overlap.

4.2. Theoretical Model

4.2.1. Fluctuations of Travel Time in Inhomogeneous Air

Travel Time Integral for Small Inhomogeneities

The high-frequency approximation of the propagation time of plane waves in an inhomogeneous medium, eq. (1.6), can be derived from the Eikonal [1], [4], [130],

$$t = \int_{s_0}^{s_1} \frac{1}{c(\vec{r})} ds, \qquad (4.1)$$

where $\vec{r}$ describes the coordinates along the propagation path s from s_0 to s_1 and $c(\vec{r})$ is the local speed of sound. It can be assumed that the speed of sound varies only slightly throughout the space and therefore set $c(\vec{r}) = c_0 + \tilde{c}(\vec{r})$ with constant c_0 and $\tilde{c}(\vec{r})/c(\vec{r}) \ll 1$. Typically one may specify c_0 by the condition $c_0 = \langle c(\vec{r}) \rangle$, i.e., $\langle \tilde{c}(\vec{r}) \rangle = 0$ for the stochastic ensemble average. Linearizing the integral

$$t = \int_{s_0}^{s_1} \frac{1}{c_0 + \tilde{c}(\vec{r})} ds \tag{4.2}$$

yields

$$t \approx \int_{s_0}^{s_1} \frac{1}{c_0} \big(1 - \tilde{c}(\vec{r})/c_0 \big) ds. \tag{4.3}$$

The propagation time can thus be separated into a constant part and a term that depends on the spatial fluctuations,

$$t \approx \frac{s_1 - s_0}{c_0} - \frac{1}{c_0^2} \int_{s_0}^{s_1} \tilde{c}(\vec{r}) ds. \tag{4.4}$$

Obviously one can normalize the arrival time to the constant part, $t \rightarrow t + (s_1 - s_0)/c_0$. Additionally it can be set $-\tilde{c}(\vec{r})/c_0^2 = \alpha T(s)$, where $T(s)$ is a measurable fluctuating environmental parameter and α is a related constant, so that finally there is a simple starting point:

$$t = \alpha \int_{s_0}^{s_1} T(s) ds. \tag{4.5}$$

Note that at this time the nature of the spatial variations has not been specified. Influences on the propagation of sound, such as temperature variations or air flow, can be modeled in this way as long as the fluctuations are small and the high-frequency limit of the Eikonal holds. It was also assumed that the propagation path itself will not change significantly even though there are inhomogeneities in the medium.

Correlated Fluctuations

Considering the fluctuations along the propagation path, one has to assume that fluctuations at different points of the path are correlated. In the framework of statistical theory [138], [139], [140], these can be approximated by the spatial correlation function,

$$\Big\langle \big(T(x) - \langle T \rangle \big) \big(T(y) - \langle T \rangle \big) \Big\rangle = \frac{D}{\tau} \exp \Big(-\frac{|x - y|}{\tau} \Big), \tag{4.6}$$

where τ determines the correlation length, D is the (squared) amplitude of the fluctuations and x, y are the coordinates along the propagation path. The mean $\langle ... \rangle$ is understood as the average over the stochastic ensemble of all possible realizations of the field T. One can rewrite eq. (4.6) to:

$$\langle T(x)T(y)\rangle = \langle T\rangle^2 + \frac{D}{\tau}\exp\left(-\frac{|x-y|}{\tau}\right). \tag{4.7}$$

Computing the squared standard deviation, $\sigma_T^2 = \langle T^2\rangle - \langle T\rangle^2$, in $x = y$ yields $D = \sigma_T^2\tau$ so that

$$\langle T(x)T(y)\rangle = \langle T\rangle^2 + \sigma_T^2\exp\left(-\frac{|x-y|}{\tau}\right). \tag{4.8}$$

The spatial correlation function shows an exponential decay over distance with its amplitude defined by the standard deviation of the observed quantity T.

Variation of the Arrival Time

Since the fluctuations of air can be described only statistically and not in a deterministic way, the arrival time has to be formulated in a statistical manner as well. Its squared standard deviation is given by:

$$\sigma_t^2 = \langle t^2\rangle - \langle t\rangle^2. \tag{4.9}$$

Both terms can be resolved by applying the assumption made about the spatial correlation to the integral equation of the arrival time, eq. (4.5). For the first term one has

$$\langle t^2\rangle = \alpha^2\left\langle \int_{s_0}^{s_1} T(s)ds \int_{s_0}^{s_1} T(s')ds'\right\rangle. \tag{4.10}$$

The integrals are independent so one may mix the integrands. And one may also change to relative coordinates, so that one can integrate s' relative to any given s so that $s' = s + \rho$ and $d\rho = ds'$. One obtains

$$\langle t^2\rangle = \alpha^2\left\langle \int_{s_0}^{s_1}\int_{s_0-s}^{s_1-s} T(s)T(s+\rho)d\rho ds\right\rangle. \tag{4.11}$$

Next the integrals and the ensemble average are exchanged, because the propagation path does not depend on the fluctuations, approximately. This yields

$$\langle t^2\rangle = \alpha^2\int_{s_0}^{s_1}\int_{s_0-s}^{s_1-s}\langle T(s)T(s+\rho)\rangle d\rho ds. \tag{4.12}$$

Now one can insert the definition of the spatial correlation function, eq. (4.8), and finds:

$$\langle t^2\rangle = \alpha^2\langle T\rangle^2(s_1 - s_0)^2 + \alpha^2\sigma_T^2\int_{s_0}^{s_1}\int_{s_0-s}^{s_1-s}\exp\left(-\frac{|\rho|}{\tau}\right)d\rho ds. \tag{4.13}$$

The double integral can be solved in a straight-forward manner. The result is (note $s_1 \geq s_0$):

$$\langle t^2 \rangle = \alpha^2 \langle T \rangle^2 (s_1 - s_0)^2 + 2\alpha^2 \sigma_T^2 \tau \left\{ (s_1 - s_0) + \tau \left[\exp\left(-\frac{s_1 - s_0}{\tau} \right) - 1 \right] \right\}. \qquad (4.14)$$

The second term in equation (4.9) is easily calculated:

$$\langle t \rangle^2 = \alpha^2 \langle T \rangle^2 (s_1 - s_0)^2 \qquad (4.15)$$

Finally one obtains for the squared standard deviation of the arrival time:

$$\sigma_t^2 = 2\alpha^2 \sigma_T^2 \tau \left\{ (s_1 - s_0) + \tau \left[\exp\left(-\frac{s_1 - s_0}{\tau} \right) - 1 \right] \right\}. \qquad (4.16)$$

This equation for the *travel time variance* represents a main result of this chapter. It provides the quantitative dependency of the fluctuations of the arrival time on the statistical properties of the observation variable T, such as air flow or temperature, as well as on the distance between source and receiver.

Asymptotic Limits

For long propagation paths $s_1 - s_0$ relative to the correlation length τ, the second term of the result (4.16) can be neglected compared to the first. Then the squared standard deviation will be linearly proportional to the receive distance. Evidently, the correlation over relatively short path lengths is negligible in this case:

$$\sigma_t^2 \rightarrow 2\alpha^2 \sigma_T^2 \tau (s_1 - s_0). \qquad (4.17)$$

Vice versa, for very short propagation paths compared to the correlation length, the exponential function has a very small exponent and can be expanded into a power series. The absolute and the linear term cancel, so that the second-order term remains. As a result the squared standard deviation grows quadratically in the short range and is approximately independent from the correlation length:

$$\sigma_t^2 \rightarrow \alpha^2 \sigma_T^2 (s_1 - s_0)^2. \qquad (4.18)$$

For the particular case that correlation length and receive distance are of the same order of magnitude, one can estimate convenient upper and lower bounds of the travel time variance, eq. (4.16),

$$0 \leq \sigma_t^2 \leq 2\alpha^2 \sigma_T^2 \frac{\tau (s_1 - s_0)^2}{\tau + s_1 - s_0}, \qquad (4.19)$$

from the inequality $1/(1 + x) \geq \exp(-x) \geq 1 - x$, valid for $x \geq -1$. See Fig. 4.1 for a graphical display of the travel time variance, its asymptotic limits, eqs. (4.17) and (4.18), as well as its upper bound, eq. (4.19). It is visible that already at distances slightly larger than the correlation length, eqs. (4.17) and (4.19) provide fair practical proxies, also suitable to simplify exact results derived below, such as eq. (4.57) in Section 4.5.

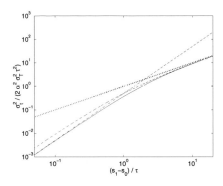

Figure 4.1.: Normalized travel time variance, exact $(-)$ according to eq. (4.16), asymptotic limit for long distances $(\cdots)$, $(s_1 - s_0)/\tau \to \infty$, according to eq. (4.17), asymptotic limit for short distances $(-.-)$, $(s_1 - s_0)/\tau \to 0$, according to eq. (4.18) and upper bound $(- -)$ according to eq. (4.19).

Examples

When considering spatial fluctuations of the speed of sound due to air flow or wind one has $\alpha = -1/c_0^2$, $T(s) = v(\vec{r}(s))$ and $\sigma_T = \sigma_v$, where v is the local wind velocity and σ_v is the standard deviation of the wind velocity. In the long-range limit of $s_1 - s_0 \gg \tau$ one obtains for the spreading of arrival times:

$$\sigma_t \approx \left|\frac{1}{c_0^2}\right| \sigma_v \sqrt{2\tau(s_1 - s_0)}. \qquad (4.20)$$

As an example, set $c_0 = 340$ m/s, $\sigma_v = 1$ m/s, $\tau = 1$ m and $s_1 - s_0 = 100$ m. This yields $\sigma_t = 0.122$ ms, which corresponds to slightly more than one wavelength at 10 kHz.

For spatial fluctuations of the Celsius temperature θ and their effect on the speed of sound [1] it is known:

$$c(\vec{r}) \approx c_0 \sqrt{1 + \theta(\vec{r})/\theta_0}. \qquad (4.21)$$

with $\theta_0 = 273.15$ °C. Linearizing for small θ/θ_0 in the definition of $1/c$ yields

$$\frac{1}{c(\vec{r})} \approx \frac{1}{c_0} - \frac{\theta(\vec{r})}{2c_0\theta_0}, \qquad (4.22)$$

so that $\alpha = -1/(2c_0\theta_0)$, $T(s) = \theta(\vec{r}(s))$ and $\sigma_T = \sigma_\theta$. Thus one finds for the arrival time in the long-range limit:

$$\sigma_t \approx \left|\frac{1}{2c_0\theta_0}\right| \sigma_\theta \sqrt{2\tau(s_1 - s_0)}. \qquad (4.23)$$

As an example, set $c_0 = 340$ m/s, $\sigma_\theta = 1$ K, $\tau = 1$ m and $s_1 - s_0 = 100$ m. This yields $\sigma_t = 0.076$ ms, which corresponds to slightly less than one wavelength at 10 kHz.

4.2.2. Fluctuations of the Travel Time Difference on Coupled Paths

When discussing the loss of coherence between two signals caused by travel time fluctuations one must account for spatial correlations between the corresponding propagation paths. Following the line of thought from the previous section the (fluctuating) difference Δt of the arrival times along these two paths are of particular interest. The squared standard deviation of this difference is given by

$$\sigma_{\Delta t}^2 = \langle \Delta t^2 \rangle - \langle \Delta t \rangle^2. \tag{4.24}$$

One may define t_1 and t_2 as the arrival time along each propagation path, respectively. This means $\Delta t = t_1 - t_2$, so that

$$\sigma_{\Delta t}^2 = \langle t_1^2 \rangle - \langle t_1 \rangle^2 + \langle t_2^2 \rangle - \langle t_2 \rangle^2 + 2\langle t_1 \rangle \langle t_2 \rangle - 2\langle t_1 t_2 \rangle. \tag{4.25}$$

Since the auto-correlation function was already computed as well as the mean of independent paths in eqs. (4.14) and (4.15) the only unknown term in this equation is the covariance term $\langle t_1 t_2 \rangle$. Applying definition (4.9) for each propagation path results in

$$\sigma_{\Delta t}^2 = \sigma_{t_1}^2 + \sigma_{t_2}^2 + 2\alpha^2 \langle T \rangle^2 (s_1 - s_0)(s_2 - s_0) - 2\langle t_1 t_2 \rangle. \tag{4.26}$$

In order to compute $\langle t_1 t_2 \rangle$ one has to determine the path integral (4.5) separately for each propagation path, $s_0 \to s_1$ and $s_0 \to s_2$. For the sake of simplicity, in the following the receiver location will be chosen as $s_0 = 0$. This yields

$$\langle t_1 t_2 \rangle = \alpha^2 \langle \int_0^{s_1} T(s)ds \int_0^{s_2} T(s')ds' \rangle. \tag{4.27}$$

In this case the correlation function must be computed for two points on different paths so that one obtains the spatial correlation function of eq. (4.8) in a generalized form,

$$\langle T(x)T(y) \rangle = \langle T \rangle^2 + \sigma_T^2 \exp\left(- \frac{r(x,y)}{\tau} \right), \tag{4.28}$$

where $r \geq 0$ is the distance between a particular point x on path s and another point y on path s'. Accordingly, the problem that finally needs to be solved can be written as

$$\sigma_{\Delta t}^2 = \sigma_{t_1}^2 + \sigma_{t_2}^2 - 2\alpha^2 \sigma_T^2 \int_0^{s_1} ds \int_0^{s_2} ds' \exp\left(- \frac{r(s,s')}{\tau} \right), \tag{4.29}$$

where

$$r = \sqrt{s^2 + s'^2 - 2ss' \cos\gamma}, \tag{4.30}$$

and $180° \geq \gamma \geq 0°$ is the angle enclosed between the two straight paths which intersect at the receiver location, $s = s' = 0$.

Notice that here a joint starting point and different end points have been chosen. This is without loss of generality because the end result of the integral term in eq. (4.29) remains the same when both integration boundaries are inverted at the same time and thus switch to a joint end point and different starting points.

Approximation and Upper Bound of the Correlation Integral

The integral that needs to be solved for two correlated paths is:

$$I = \int\limits_{0}^{s_1} ds \int\limits_{0}^{s_2} ds' e^{-r(s,s')/\tau}. \tag{4.31}$$

Looking at the exponent according to eq. (4.30) it seems useful to introduce the ratio $u = s/s'$ so that $r = s'\sqrt{u^2 + 1 - 2u\cos\gamma}$. In the case $u \gg 1$, or $s \gg s'$ respectively, one can approximate eq. (4.30) in first-order terms by

$$r_{(1)} \approx s - s'\cos\gamma. \tag{4.32}$$

Since the problem is symmetrical in s and s' one finds the counterpart for $u \ll 1$, or $s' \gg s$ respectively, immediately:

$$r_{(3)} \approx s' - s\cos\gamma. \tag{4.33}$$

Finally, in the case $u \approx 1$, or $s' \approx s$ one can linearize eq. (4.30) in the form $s = s' + \Delta s$ for small Δs. Omitting terms of higher order and applying a trigonometric identity yields

$$r_{(2)} \approx (s + s')\sin(\gamma/2). \tag{4.34}$$

Note that $r \geq 0$ applies always.

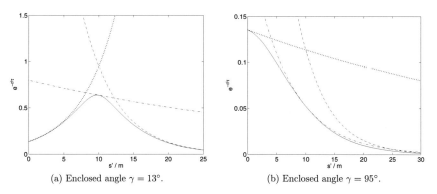

(a) Enclosed angle $\gamma = 13°$. (b) Enclosed angle $\gamma = 95°$.

Figure 4.2.: Value of exponential correlation function, that is the integrand of eq. (4.31), for $s = 10$ m and $\tau = 5$ m: exact ($-$), upper bounds for $s \gg s'$ ($\cdots$), $s' \approx s$ (-.-), and $s \ll s'$ (- -).

Interestingly these approximations also represent an upper bound of the problem. It can be shown that the approximation of r by eqs. (4.32), (4.33), or (4.34) is always smaller than or equal to the exact distance r in eq. (4.30), see Appendix A.4.1. This means that eq. (4.31) is an upper limit of the exact solution when r is replaced by one of its above approximations. Fig. 4.2 shows two typical examples where one can recognize that the combination of the three upper-bound functions represents a fairly good approximation[2]. It is noted that an upper bound for the correlation integral, eq. (4.31), corresponds to a lower bound for the variance of the travel time difference, eq. (4.29).

Since the integral in eq. (4.31) is rather difficult to solve it will be approximated instead. For any value of s the inner integral of eq. (4.31) can be subdivided into three different regions where $s' \ll s$, $s' \gg s$ or $s' \approx s$. From here on one may also assume that $s_2 \geq s_1$ without loss of generality due to the symmetry in s and s'. The transition lines between the regimes can be defined by means of the intersection points of the approximating functions. As Fig. 4.2 illustrates, this choice minimizes the excess area of the linear approximation that is added to the exact integral.

The first transition is between the regions of $s' \ll s$ and $s' \approx s$. Here the intersecting line is given by the equality of $r_{(1)}$ and $r_{(2)}$ in $s' =: s_a$:

$$s - s_a \cos\gamma = (s + s_a) \sin(\gamma/2). \tag{4.35}$$

One obtains:

$$s_a(s) = s \frac{1 - \sin(\gamma/2)}{\cos\gamma + \sin(\gamma/2)}. \tag{4.36}$$

The second transition happens between $s' \approx s$ and $s' \gg s$ in $s' =: s_b$ where $r_{(2)}$ equals $r_{(3)}$:

$$(s + s_b) \sin(\gamma/2) = s_b - s \cos\gamma, \tag{4.37}$$

yielding

$$s_b(s) = s \frac{\cos\gamma + \sin(\gamma/2)}{1 - \sin(\gamma/2)}. \tag{4.38}$$

For the sake of brevity one may introduce the slope $v = \frac{1-\sin(\gamma/2)}{\cos\gamma+\sin(\gamma/2)}$ so that $s_a(s) = sv$ and $s_b(s) = s/v$. Note that $1 \geq v \geq 1/3$ for all γ.

After these preparations the relationship (4.31) can be redefined by subdividing the inner integral into the discussed three approximation regions (see Fig. 4.3):

$$I = \int_0^{s_1} ds \left[\int_0^{s_a(s)} ds' e^{-r_{(1)}/\tau} + \int_{s_a(s)}^{s_b(s)} ds' e^{-r_{(2)}/\tau} + \int_{s_b(s)}^{s_2} ds' e^{-r_{(3)}/\tau} \right]. \tag{4.39}$$

But notice that this expression applies only over the full extent of the outer integral when $s_2 \geq s_b(s_1)$. This is not true for the typical case of $\gamma > 0$ and $s_2 = s_1$. In this complementary case of $s_b(s_1) > s_2 \geq s_1$ one must also subdivide the outer integral, as

---

[2]As far as *exact* results of the integral are given in the following, they are computed numerically using standard methods.

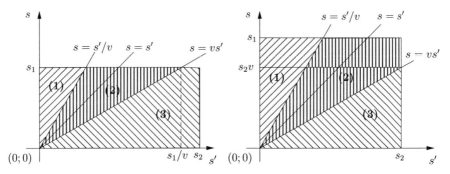

Figure 4.3.: Integration regions in the case $s_2 \geq s_1/v$.

Figure 4.4.: Integration regions in the case $s_1/v > s_2 \geq s_1$.

shown in Fig. 4.4. The case $s_2 < s_1$ does not have to be considered due to the symmetry in s and s'.

One finds accordingly for the second case with $s_b(s_1) > s_2$:

$$I = \int_0^{s_2 v} ds \left[\int_0^{s_a(s)} ds' e^{-r_{(1)}/\tau} + \int_{s_a(s)}^{s_b(s)} ds' e^{-r_{(2)}/\tau} + \int_{s_b(s)}^{s_2} ds' e^{-r_{(3)}/\tau} \right]$$

$$+ \int_{s_2 v}^{s_1} ds \left[\int_0^{s_a(s)} ds' e^{-r_{(1)}/\tau} + \int_{s_a(s)}^{s_2} ds' e^{-r_{(2)}/\tau} \right]. \quad (4.40)$$

Since all three approximations of r are linear functions of s and s' this double integral can be solved in a straight-forward manner. Notice that beyond the break-up and linearization of $r(s, s')$ for different s/s' no assumptions have been made regarding s_1, s_2, γ or τ. Therefore this upper bound is valid for all possible configurations. The result of executing the integrals is somewhat lengthy and does not provide additional insights. For the sake of completeness it is given in Appendix A.4.1.

Error of Approximation

When comparing the linear approximations (4.39), (4.40) with the exact integral (4.31) one finds that for the largest part of the parameter space the deviation is within a few percent or less (see e.g. Fig. 4.5) and does not affect the result of eq. (4.29) significantly. However, especially for small angles γ and large distances s_1 and s_2 with $s_1 = s_2$, the error of the approximation becomes important. This is due to the fact that in this case the correlated part in eq. (4.29) becomes almost as large as the uncorrelated part that it is subtracted from.

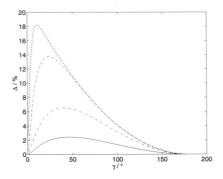

Figure 4.5.: Relative deviation of approximation (4.40) from exact integral (4.31) when $s_1 = s_2$ and $s_1 = \tau$ (−), $s_1 = 3\tau$ (-.-), $s_1 = 10\tau$ (- -), and $s_1 = 30\tau$ (···).

Unfortunately, this is a typical situation of application when considering line array systems. Here the spacing between the sources will be relatively small compared to the listening distance. At the same time the distance from each source to the receiver is almost the same. In order to demonstrate this practical application the angle γ can be formally translated into a vertical spacing $h = 2s_1 \sin(\gamma/2)$ between the two sources. It is assumed that the receiver is centered vertically between the two sources and that the distance to the sources is s_1 and s_2, respectively, with $s_1 = s_2$.

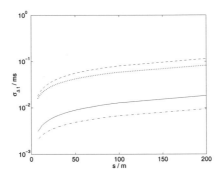

Figure 4.6.: Exact and approximate standard deviation of arrival time difference $\sigma_{\Delta t}$ for small γ and large distances $s_1 = s_2$ according to eq. (4.29). Source spacing $h = 0.3$ m, exact (−) and lower bound (-.-), spacing $h = 3$ m, exact (- -) and lower bound (···).

The resulting dependency of the standard deviation on distance and spacing is shown in Fig. 4.6 as an example. For this illustration only fluctuations due to light wind have been considered. A correlation length of $\tau = 5$ m has been assumed and a standard deviation of $\sigma_v = 0.7$ m/s. In the range of interest of 10 m $\geq h \geq 0.1$ m and

200 m $\geq s_1 \geq 10$ m, the lower-bound underestimates the accurate value by about 5% to 55%. An upper bound for the standard deviation of the propagation time difference could thus be defined empirically; it is roughly twice the value of the lower bound.

This graph allows deriving some practical numbers. According to the model, if the spacing of acoustic sources is of the order of 1 m, the standard deviation of the arrival time difference can be up to 0.1 ms. This corresponds to one wavelength at 10 kHz.

Asymptotic Limits

Equation (4.29) allows discussing a number of asymptotic cases. In the case of $\gamma \to 0$ the distance becomes simply $r = |s - s'|$. One finds:

$$\int_0^{s_1} ds \int_0^{s_2} ds' e^{-|s-s'|/\tau} = \tau^2 \left(2\frac{s_1}{\tau} + e^{-s_1/\tau} + e^{-s_2/\tau} - e^{-(s_2-s_1)/\tau} - 1 \right). \qquad (4.41)$$

Without loss of generality this solution assumes $s_2 \geq s_1$. Inserting this result as well as eq. (4.16) into the governing eq. (4.29) yields

$$\sigma_{\Delta t}^2 = 2\alpha^2 \sigma_T^2 \tau^2 \left\{ \frac{s_2 - s_1}{\tau} + \exp\left(-\frac{s_2 - s_1}{\tau} \right) - 1 \right\}. \qquad (4.42)$$

Obviously, for small angles γ the standard deviation of the travel time difference is equivalent to the uncorrelated (single-path) solution for the path length difference $s_2 - s_1$. This is because the sound waves propagating along the shared part of the path will see exactly the same fluctuations. They only differ by one propagation path extending beyond the other and thus one sound wave accumulating the effects of additional fluctuations on that excess section. As expected, the travel time difference vanishes completely when the two paths have the same length, $s_1 = s_2$.

In the case of $\gamma \to 180°$ the relation $r = s + s'$ applies. Solving the integral in eq. (4.31) is straight-forward:

$$\int_0^{s_1} ds \int_0^{s_2} ds' e^{-(s+s')/\tau} = \tau^2 \left(e^{-(s_1+s_2)/\tau} - e^{-s_2/\tau} - e^{-s_1/\tau} + 1 \right). \qquad (4.43)$$

Clearly, for $s_1 \gg \tau$ and $s_2 \gg \tau$ this integral and thus the correlated part in eq. (4.29) become very small relative to the uncorrelated part. Not unexpectedly, for large distances s_1, s_2 and large opening angles γ the standard deviation of the arrival time difference is dominated by the uncorrelated part. Comparing with eq. (4.17) this result corresponds to the uncorrelated (single-path) solution for a path length of $s_1 + s_2$.

4.2.3. Coherent Superposition

Once the fluctuations of the arrival time are determined, whether this is from theory or from experiments, one can calculate the corresponding phase fluctuations. The mean is given by the mean of the arrival time multiplied by frequency ω:

$$\langle \phi \rangle = \omega \langle t \rangle \tag{4.44}$$

The standard deviation obeys the same relationship:

$$\sigma_\phi = \omega \sigma_t \tag{4.45}$$

But naturally the phase fluctuations are not very meaningful unless their effect on the superposition of signals is considered when they take different paths to the receiver.

Coherent Superposition of Two Remote Sources

In this section an idea will be given on how the statistical variation of the signal arrival time will affect the summation of two signals at the receiver. One may consider two sources which radiate the same signal. They are located at the same distance to the receiver but in different directions, so that the fluctuations on the two paths are practically uncorrelated.

When the receiver is located in the origin the complex sound pressure p_j of a spherical wave radiated by source j at distance r_j is determined by

$$p_j = \frac{1}{r_j} A_j e^{-i(kr_j - \xi_j)}, \tag{4.46}$$

where A_j is the complex radiation amplitude of the source, k the wave number and ξ_j the instantaneous phase fluctuation relative to the phase that corresponds to the unperturbed arrival time $t_j = r_j/c_0$. For two sources, 1 and 2, with equal amplitude and at equal distance the pressure sum $p_{Sum} = p_1 + p_2$ yields after some simplifications

$$p_{Sum} = A\left(e^{i\xi_1} + e^{i\xi_2}\right), \tag{4.47}$$

where all joint quantities have been compacted into the complex amplitude A. One may calculate the power of the complex sum, $P = |p_{Sum}|^2$, and define the phase difference $\xi = \xi_1 - \xi_2$:

$$P = 2|A|^2(1 + \cos\xi). \tag{4.48}$$

For Gaussian-distributed phase fluctuations it is known that $\langle \xi \rangle = \langle \xi_1 \rangle - \langle \xi_2 \rangle$ and $\sigma_\xi^2 = \sigma_{\xi_1}^2 + \sigma_{\xi_2}^2$, [120]. Note that the second expression corresponds to the uncorrelated part in eq. (4.29).

One can now average over the ensemble by assuming a Gaussian distribution W with the mean $\langle \xi \rangle$ and the standard deviation σ_ξ:

$$\langle P \rangle = 2|A|^2\left(1 + \int_{-\infty}^{\infty} W(\xi)\cos\xi \, d\xi\right), \tag{4.49}$$

where

$$W(\xi) = \frac{1}{\sqrt{2\pi}\sigma_\xi} \exp\left(-\frac{(\langle \xi \rangle - \xi)^2}{2\sigma_\xi^2}\right). \tag{4.50}$$

The integral boundaries are infinite because the average is over all possible realizations of the fluctuations of the arrival time difference multiplied by ω. One may choose $\langle \xi \rangle = 0$ and finds after executing the integral:

$$\langle P \rangle = 2|A|^2 \left(1 + \exp \left(-\frac{\sigma_\xi^2}{2} \right) \right). \tag{4.51}$$

This result provides the transition between the fully coherent state $\sigma_\xi = 0$ when $\langle P \rangle = 4|A|^2$ and the random-phase state $\sigma_\xi \to \infty$ when $\langle P \rangle = 2|A|^2$.

As an example, assume that the standard deviation of the arrival time difference corresponding to ξ is $\sigma_t = 0.1$ ms. This yields an average power sum $\langle P \rangle = 2.0|A|^2$ at 10 kHz, $\langle P \rangle = 2.34|A|^2$ at 3 kHz, $\langle P \rangle = 3.64|A|^2$ at 1 kHz and $\langle P \rangle = 3.996|A|^2$ at 100 Hz. Obviously, for a phase variation of one cycle, the coherent part in the overall power received has vanished.

The assumption that the distribution of fluctuations over the ensemble is Gaussian is plausible if the two considered propagation paths are independent and do not show spatial or temporal correlations. In practice, this will only be the case if - looking from the receiver - the relative angle between these paths is sufficiently large, that is, if the sources are spaced much further apart than the correlation distance.

Coherent Superposition of Spatially Close Sources

Equation (4.29) represents the standard deviation of the travel time difference for two correlated paths. It includes the special case of uncorrelated paths, eq. (4.16), in the asymptotic limit $\gamma \to 180°$ and $s_1 \gg \tau$, $s_2 \gg \tau$. With respect to the superposition of multiple sources one can draw conclusions similar to the above combination of sound waves traveling along uncorrelated paths and generalize the result. The standard deviation of the fluctuating phase difference between two sources is then given by $\sigma_\xi = \omega \sigma_{\Delta t}$ and relationship (4.51) applies unchanged.

When considering more than two sources the phase relationship must be determined for each pair of sources individually. Let N sources j have the same amplitude A and the instantaneous phase ξ_j at the receive location. The sum is established by

$$p_{Sum} = \sum_{j=1}^{N} A e^{i\xi_j}. \tag{4.52}$$

The power of the complex sum is given by

$$P = |p_{Sum}|^2 = |A|^2 \left(N + 2 \sum_{j=1}^{N} \sum_{k=j+1}^{N} \cos \xi_{j,k} \right), \tag{4.53}$$

where $\xi_{j,k} = \xi_j - \xi_k$ is the instantaneous phase difference between two sources j and k. Derivation of the mean power sum by applying the same principle as for eq. (4.51) is straight forward under the assumption that the distribution function of the travel time difference is roughly Gaussian:

$$\langle P \rangle = |A|^2 \left(N + 2 \sum_{j=1}^{N} \sum_{k=j+1}^{N} \exp\left(-\frac{\sigma_{\xi_{j,k}}^2}{2} \right) \right). \tag{4.54}$$

Finally one obtains:

$$\langle P \rangle = |A|^2 \left(N + 2 \sum_{j=1}^{N} \sum_{k=j+1}^{N} \exp\left(-\frac{\omega^2 \sigma_{\Delta t_{j,k}}^2}{2} \right) \right), \tag{4.55}$$

where $\sigma_{\Delta t_{j,k}}$ is the standard deviation of the difference of the propagation times along two paths s_j and s_k. In combination with eq. (4.29) this is an important result of this chapter.

Practical implications of these theoretical results will be discussed in Section 4.5.

4.3. Experimental Results

Under calm summerly conditions in a suburb region with single houses, bushes, and meadows, outdoor measurements were carried out on two subsequent days. The measurement setup consisted of two anemometers which were located at a specific distance from each other. 29 time data sets, regarded as legs in this chapter, were obtained covering measurement periods from 1 min to 30 min (see Appendix A.4).

Environmental parameters were measured using ultrasonic anemometers, a Thies Ultrasonic Anemometer 3D [141] and a Gill WindObserver II [142]. The instruments are depicted in Figs. A.6 and A.7, respectively, in Appendix A.4. The anemometer measurements include in particular wind direction and velocity in the horizontal plane as well as the air temperature. Data samples were acquired at rates of 10 Hz and 7 Hz respectively, each with an averaging period of 100 ms. This resolution should provide a reasonable signal-to-noise ratio while keeping the data rate fast enough for relevant changes of air flow and temperature.

Later, also acoustic impulse response measurements will be looked at. These were performed using one or two loudspeakers and a measurement microphone. Loudspeakers and microphone were mounted about 1.5 m above the ground. The propagation axis from loudspeaker to microphone was also chosen in the main wind direction.

4.3.1. Wind Velocity

Anemometer measurement data for the wind speed and direction were processed to *longitudinal* wind velocity and *transversal* wind velocity. In this study, instantaneous wind vectors are defined as *longitudinal*, when they point in the main wind direction, and *transversal* when they are orthogonal to the longitudinal direction.

Time Dependence

Fig. 4.7a shows a typical measurement of the longitudinal wind velocity over a period of 600 s. Fig. 4.7b shows the same data in a magnified view for the section of 120 s to 180 s. The corresponding transversal wind velocity is displayed in Figs. 4.8a and 4.8b. The distribution functions of longitudinal and transversal wind velocity are depicted in Figs. 4.7c and 4.8c, respectively.

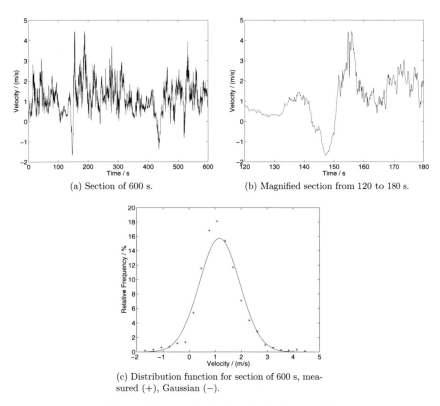

(a) Section of 600 s.

(b) Magnified section from 120 to 180 s.

(c) Distribution function for section of 600 s, measured ($+$), Gaussian ($-$).

Figure 4.7.: Longitudinal wind velocity, leg 10.

The diagrams show wind speeds of up to 4 m/s in the longitudinal direction and ± 1.5 m/s in the transversal direction. The longitudinal data are largely greater than zero which should be the case if they are measured in the main wind direction.

It should be noted that the data exhibit events on two different time scales. Quick changes of relatively small amplitude, from sample-to-sample (~ 0.1 s) define the fast

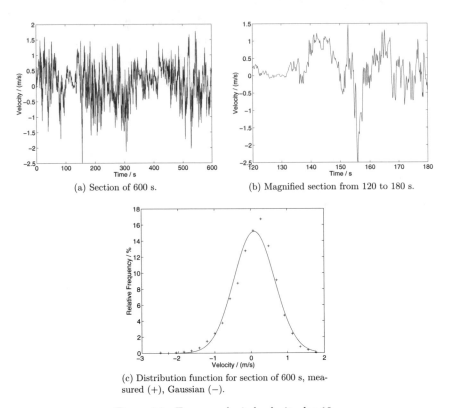

(a) Section of 600 s.

(b) Magnified section from 120 to 180 s.

(c) Distribution function for section of 600 s, measured (+), Gaussian (−).

Figure 4.8.: Transversal wind velocity, leg 10.

time scale. These are related to microscopic turbulence in the air. Possible signals at higher rates than 0.1 s are not resolved by the instruments. Changes of higher amplitude take place at a slower time scale (∼ 5-10 s) and represent the actual macroscopic air flow. For either longitudinal or transversal data the pictures are similar.

A direct overlay of longitudinal and transversal data for this example is given in Fig. 4.9. It is interesting that for some time sections the longitudinal and the (negative) transversal velocity are correlated which hints at oblique incidence angles. Together with the phase-shifted rapid inversion of the flow direction, this signal indicates the passage of a turbulent eddy.

Already from this example one can derive that typical correlation times for the air flow will be of the order of 5 s.

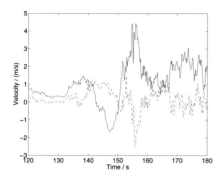

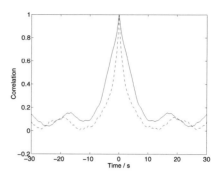

Figure 4.9.: Comparison of longitudinal (−) and transversal (- -) data, leg 10, magnified section from 120 to 180 s.

Figure 4.10.: Auto-correlation functions of longitudinal (−) and transversal (- -) data, leg 10, for the time-lag section from -30 to 30 s.

The standard deviation of the time domain data can be used to quantify the average amplitude of the fluctuations. In this particular example a standard deviation of 0.77 m/s is found for the longitudinal direction and 0.55 m/s for the transversal direction. For all experiments, the standard deviation is between 0.4 m/s and 1.0 m/s. Table A.2 lists the standard deviations of all measurements taken for both anemometers. The different values for the standard deviation are of the same order of magnitude. They do not show a significant dependence on the measurement setup. The same applies to the distribution functions of the acquired data; they can be regarded as Gaussian, approximately.

Correlation Functions

In order to quantify the degree of correlation in the time domain the normalized auto-correlation function of the data set can be calculated. Fig. 4.10 shows the auto-correlation plots for the above longitudinal and transversal example. Clearly, correlation times are in the range of 5 s. Other acquired data sets show comparable behavior.

This finding supports the previous reasoning. It means that at any point of time when one looks at the propagation of a sound signal, the environmental conditions along the propagation path can be safely assumed to be time-invariant as long as the travel times are significantly smaller than the correlation time. For a correlation time of 5 s this translates to travel distances of less than 1.6 km.

For the purpose of investigating the spatial properties of the wind field Figs. 4.11a and 4.11b provide overlays of two longitudinal measurement sets, acquired at the same time but at different locations. In this case, measurement points are spaced 2.4 m apart. The similarity of such data as a function of distance between the measurement locations

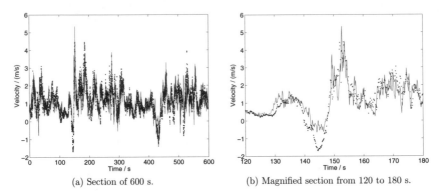

(a) Section of 600 s.

(b) Magnified section from 120 to 180 s.

Figure 4.11.: Comparison of longitudinal data sets for two measurement locations at 2.4 m distance, leg 10, Gill ($-$), Thies ($\cdots$).

allows deriving information about the spatial correlation of the wind velocity. This example shows a high correlation but also some clear deviations between the longitudinal measurements.

A good way to measure the spatial correlation is to compute the cross-correlation function of the two data sets. Fig. 4.12 shows the normalized Pearson cross-correlation function for both the longitudinal and the transversal wind velocity. Basically its maximum value indicates the degree of correlation between the wind fluctuations at the two measurement points, allowing for some time lag due to the advection of air between the measuring sites.

For this example, both longitudinal and transversal measurements show a distinct maximum indicating a fairly high degree of correlation and a clear decay as well. The cross-correlation coefficient is 0.72 and 0.66 respectively.

Figure 4.13 shows the maximum value of the cross-correlation function for a number of measurements of both longitudinal and transversal wind velocity. The complete data are listed in Appendix A.4, Table A.3. Visibly, the correlation decreases with distance. The logarithmic plot in Fig. 4.13 shows good agreement with the assumption of an exponential correlation function. The fit yields a correlation length of 5.9 m for the longitudinal data and a correlation length of 4.4 m for the transversal data. As it seems, for the considered wind speeds typical correlation lengths are of the order of 5 m. At distances beyond this value, one measurement location will not show a significant similarity to another measurement location.

It can be recognized that the correlation coefficient does not depend significantly on the wind direction. The values for the longitudinal direction show a slightly stronger correlation than the transversal direction, as one would expect. But the difference is within the measurement uncertainty. It can be summarized that the spatial correlations are approximately isotropic in the horizontal plane.

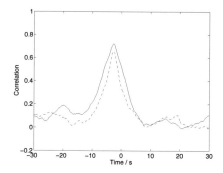 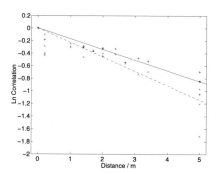

Figure 4.12.: Cross-correlation functions of longitudinal ($-$) and transversal (- -) data for two measurement locations at 2.4 m distance, leg 10, for the time-lag section from -30 to 30 s.

Figure 4.13.: Natural logarithm of longitudinal ($+$) and transversal ($\times$) cross-correlation coefficient, exponential fit function for longitudinal ($-$) and transversal coefficients (- -), as a function of the distance between measurement locations.

All measurements considered in Table A.3 and in Fig. 4.13 used measurement periods of at least 10 min to allow for a high signal-to-noise ratio.

Practical Remarks

It should be remarked that for the observations only two anemometers were available. Ideally, one would measure simultaneously at more locations at the same time to determine the spatial correlation. Practically, assuming that the statistical properties of the wind field are steady over a longer period of time, measurements at different distances were performed subsequently and are expected here to produce consistent results.

Furthermore, measurements were performed at average wind speeds of 2-4 m/s. This corresponds to a light to gentle breeze (Beaufort 2-3) which is a typical condition for outdoor events. One may assume that the statistical properties of the wind field will not change significantly at slightly higher or lower wind speeds. But for much higher wind speeds the correlation of the wind velocity will become anisotropic and the standard deviations will increase as well.

It should also be noted that measurements were made in a yard, roughly in the horizontal domain at 1-2 m above ground. It is likely that the findings presented here are approximately valid for the vertical domain near ground and for practically interesting altitudes of 10-20 m, as well [61].

4.3.2. Temperature

The two ultrasonic anemometers facilitate the determination of the so-called virtual acoustic temperature. The method they use exploits the fact that temperature changes in air will change the propagation speed of sound in an isotropic manner, in contrast to air flow. From comparing the sound speed between opposite directions, with every measurement of the wind vector the virtual acoustic temperature was also acquired.

During all measurement series the virtual acoustic temperature seemed to drift on fairly long time scales ($\sim$ 1 min). Part of the reason was certainly that temperature changes took place in the air. Other effects may include sunlight on the ultrasonic sensors themselves and its reflection by the housing of the device. However, for the considerations here absolute quantities are irrelevant and the short-term fluctuations appeared to be reliable. Accordingly, before further post-processing all temperature data sets were high-pass filtered at around 0.02 Hz and thus cleaned from typical, possibly spurious low-frequency signals.

Time Dependence

Figs. 4.14a and 4.14b show the processed measurement data of the virtual acoustic temperature, Fig. 4.14c the corresponding distribution function. As an example, the raw data are shown in Figs. A.8a and A.8b in Appendix A.4. At an absolute temperature of about 18 °C, maximum variations of the temperature are in the range of 0.5 K. The standard deviation in this example is about 0.1 K.

For all experiments, the standard deviation is not less than 0.08 K and not greater than 0.23 K. Table A.4 in Appendix A.4 shows the standard deviation for all temperature measurements. The associated distribution functions are approximately Gaussian.

Correlation Functions

Compared to the wind vector, the temperature data show shorter correlation times and shorter correlation distances. The example data used so far were acquired at a distance of 2.4 m. At this spacing between measurement sites the temperature data exhibits no clear correlation anymore. Therefore an example with a distance of 1.4 m is used to illustrate the correlation properties. Figs. 4.15a and 4.15b show the corresponding time data as an overlay. Figure 4.16 presents the auto-correlation functions for the two locations as well as the cross-correlation function.

Fig. 4.17 displays the maximum value of the cross-correlation function for a number of measurement setups. Table A.5 in Appendix A.4 gives an overview over all measurements.

The auto-correlation plots show a clear and quick decrease of correlation over time. One can estimate a correlation time of about 1-2 s. The cross-correlation of temperature over distance also decreases more strongly than that of the wind velocity and the data are noisier. The fit function for the correlation length yields 2.25 m, which is roughly 2 m.

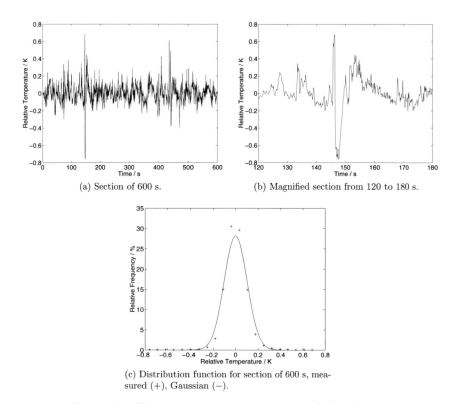

(a) Section of 600 s.

(b) Magnified section from 120 to 180 s.

(c) Distribution function for section of 600 s, measured (+), Gaussian (−).

Figure 4.14.: Temperature measurement, processed data, leg 10.

The acquired temperature data show fluctuations with amplitudes of approximately 0.15 K. Because a change of 0.15 K in temperature corresponds to a change of about 0.1 m/s in wind velocity at the given measurement conditions, see eq. (4.21), one can safely assume that the effect of temperature variations on the propagation of sound will be significantly smaller than that of fluctuations in the air flow. In addition, the relatively short correlation distances will also reduce the influence of the temperature changes. In the following the focus will be therefore on the relationship between wind velocity and the propagation delay of sound.

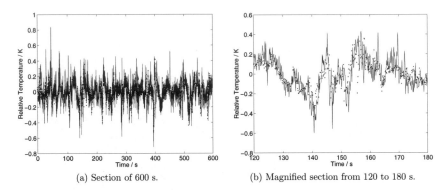

(a) Section of 600 s.   (b) Magnified section from 120 to 180 s.

Figure 4.15.: Comparison of temperature data sets for two measurement locations at 1.4 m distance, leg 22, Gill ($-$), Thies ($\cdots$).

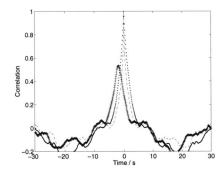

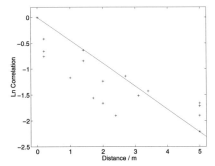

Figure 4.16.: Auto-correlation (- -), ($\cdots$) and cross-correlation ($+$) functions of temperature data for two measurement locations at 1.4 m distance, leg 22, for the time-lag section from -30 to 30 s.

Figure 4.17.: Natural logarithm of cross-correlation coefficient ($+$), exponential fit function ($-$), as a function of the distance between measurement locations.

4.3.3. Propagation of Sound

The second part of the measurements consisted of the acquisition of acoustic impulse responses in order to determine the propagation delay along the longitudinal axis. Acoustic measurements were performed with the FFT-based measurement platform EASERA [143]. A short sweep signal of 42 ms was used to provide for a measurement rate that is high enough to capture the fluctuations of the air flow while achieving sufficient signal-to-noise ratio for the calculation of the arrival time.

Four measurement series were carried out with a single loudspeaker and a microphone (legs 7-10). Another four measurement series were made with two loudspeakers at different base angles (legs 21-24). Legs 21 and 22 used a small base angle of about 13° and leg 23 used an angle of about 25°. In leg 24 loudspeakers and anemometers and microphone were all aligned (see also Appendix A.4). Each measurement series consisted of 1200 measurements, covering approximately 10 min of time overall. When using one loudspeaker, the two anemometers were both located on the measurement axis. When using two loudspeakers, the loudspeakers were aimed at the microphone and each anemometer was located on one measurement axis. The acquired impulse response data were scanned and the arrival time of the sound signal was retrieved. For the setup with two loudspeakers, the propagation time was determined separately for each loudspeaker.

Propagation Time

Analyzing the propagation delay as a function of measurement time provides a function like the example displayed in Fig. 4.18a. Fig. 4.18b shows the distribution function of the same data. In this case the measurement distance was approximately 9.2 m which results in a time of flight of about 27 ms. Fluctuations about this value are due to changes of the environmental parameters along the propagation path (assuming stable microphone and loudspeaker performance).

In contrast to the wind velocity measurements discussed in the previous section, the fast fluctuations of small amplitude are largely absent in the travel time. This is due to the fact that according to eq. (4.1) the sound signal physically averages over all random fluctuations or inhomogeneities along the propagation path. Only those effects remain that can be considered as constant or systematic over the time period of sound propagation.

In order to compare with the theoretical discussion of the travel time variance in the foregoing section, eq. (4.16), it is of interest to investigate the standard deviation of the measurement series. Table 4.1 shows the standard deviation of the arrival time for all measurements. Roughly, the values range from 30 μs to 50 μs and each series exhibits a nearly Gaussian distribution function. The table also shows deviations between measurement series for the same distances taken at different points of time. These are due to changes in the environmental parameters.

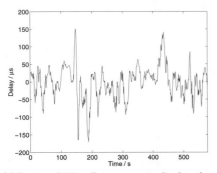 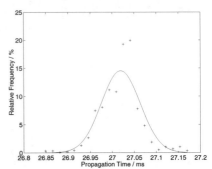

(a) Section of 600 s. Data were normalized to the mean of 27.02 ms.

(b) Distribution function for section of 600 s, measured (+), Gaussian (−).

Figure 4.18.: Propagation delay, leg 10.

Dependence on Wind Velocity

Of course, the velocity of the wind also affects the speed of sound relative to a resting coordinate frame. To confirm the relationship of wind velocity and propagation time the longitudinal velocity amplitude can be compared with the arrival time for any measurement series. Like in Figs. 4.19a and 4.19b there will be a similarity of the wind velocity measured at one point on the measurement axis and the propagation time measured along the axis. Naturally, the correlation will be smaller if the propagation distance becomes longer. In that case, the measurement point is simply less representative for the entire path, depending on the correlation length. The correlation of the amplitudes for wind velocities orthogonal to the propagation axis is irrelevant as those will affect the speed of sound only marginally (Figs. 4.20a, 4.20b).

Figures 4.21a and 4.21b show an overlay of the temperature data with the propagation time. Obviously, the temperature of the propagation medium air will affect the propagation speed and thus the arrival time. However, in this case it seems that the influence of the temperature is a second-order effect. Comparison with Figures 4.19a and 4.19b indicates that visible correlations occur only when longitudinal wind components also correlate to temperature.

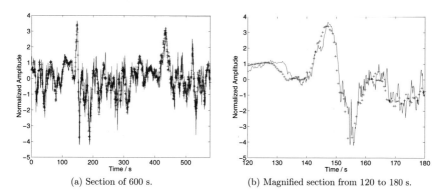

(a) Section of 600 s. (b) Magnified section from 120 to 180 s.

Figure 4.19.: Propagation time (+) and longitudinal wind velocity, Thies (−), normalized, leg 10.

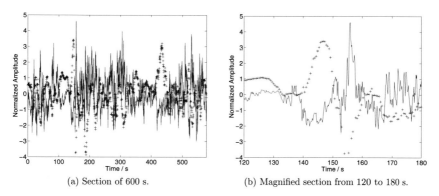

(a) Section of 600 s. (b) Magnified section from 120 to 180 s.

Figure 4.20.: Propagation time (+) and transversal wind velocity, Thies (−), normalized, leg 10.

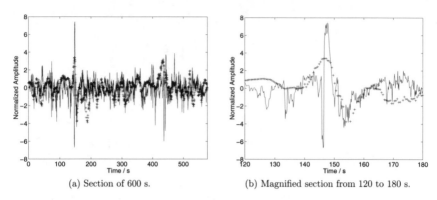

(a) Section of 600 s.

(b) Magnified section from 120 to 180 s.

Figure 4.21.: Propagation time (+) and temperature, Thies (−), normalized, leg 10.

Propagation Time Difference

After investigating the variation of the propagation time along a path and its relationship
to wind and temperature the difference between travel times on different propagation
paths should be discussed. In legs 21 to 24 propagation delays were measured for two
loudspeakers simultaneously. Since the sound waves radiated by each loudspeaker take
a slightly different path one would expect that the fluctuations of the travel times are
similar but not identical. The differences will depend on the length of the propagation
paths and on the enclosed angle.

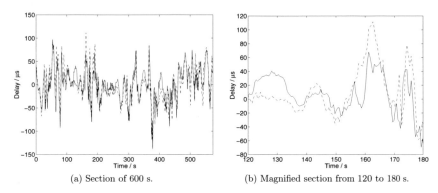

| (a) Section of 600 s. | (b) Magnified section from 120 to 180 s. |

Figure 4.22.: Propagation time of signal from loudspeaker 1 (−) and from loudspeaker
2 (- -), leg 21. Data were normalized to the respective mean, 23.87 ms for loudspeaker
1 and 29.25 ms for loudspeaker 2.

Fig. 4.22 shows the propagation delays of leg 21. They clearly exhibit a significant
degree of correlation. The difference of the mean travel times of 5.4 ms corresponds
to the difference of the travel distances of 1.8 m. Fig. 4.23 displays the differential
propagation times for the same measurement. In order to validate the theory developed
earlier the main concern is the standard deviation of the travel time difference. These
data are listed in Table 4.2 for each of the measurement series, legs 21 to 24.

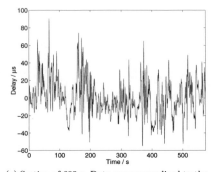

(a) Section of 600 s. Data were normalized to the mean of 5.38 ms.

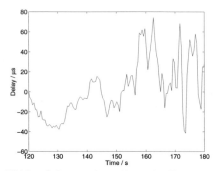

(b) Magnified section from 120 to 180 s. Data were normalized to the mean of 5.38 ms.

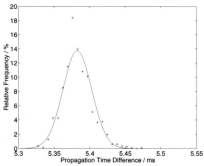

(c) Distribution function for section of 600 s, measured (+), Gaussian (−).

Figure 4.23.: Differential propagation time between signals from loudspeaker 1 and from loudspeaker 2, leg 21.

4.4. Validation

4.4.1. Propagation Time

Summing up the experimental results it can be stated that measurement data for the standard deviation of the arrival time as well as of the wind velocity and temperature have been acquired. Along with the measured correlation lengths those can now be used to validate equation (4.16) with respect to the travel time variance σ_t.

Table 4.1 compares the measured standard deviation of the propagation time with that of the data predicted using the measured wind data and temperature data[3]. Figure 4.24a shows the same data graphically, but limited to the wind velocity part.

Leg -Lsp	Path Length	Measured Std. Dev. All	Modeled Std. Dev. Velo. only	Modeled Uncertainty Velo. only	Modeled Std. Dev. Temp. only	Modeled Std. Dev. Combined
	s/m	$\sigma_t/\mu s$	$\sigma_t/\mu s$	$\delta\sigma_t/\mu s$	$\sigma_t/\mu s$	$\sigma_t/\mu s$
7	9.20	32.6	27.0	9	4.3	27.3
8	9.20	37.7	36.8	9	4.3	37.1
9	9.20	40.7	43.1	10	4.3	43.3
10	9.20	44.0	43.6	10	4.3	43.8
21-1	8.00	32.5	40.5	9	4.0	40.7
21-2	9.80	35.7	45.7	10	4.5	45.9
22-1	8.00	42.0	45.6	9	4.0	45.7
22-2	9.80	43.9	47.7	10	4.5	48.0
23-1	8.00	40.0	40.9	9	4.0	41.1
23-2	8.90	35.1	36.6	9	4.2	36.8
24-1	8.75	39.7	45.2	10	4.2	45.3
24-2	14.35	52.2	65.2	14	5.7	65.4

Table 4.1.: Measured and modeled standard deviation of propagation time for various measurement setups. See also Fig. 4.24a.

For the temperature data, a global value of $\sigma_\theta = 0.15$ K for the standard deviation was used (Table A.4). For the wind data, the individual longitudinal standard deviations σ_v for each measurement axis were used (Table A.2). For Legs 7-10 and 24, where both anemometers were located on the propagation axis, the average value was taken. The correlation lengths for the wind velocity and the temperature were assumed to be $\tau_v = 5$ m and $\tau_\theta = 2$ m, respectively. A constant sound speed of $c_0 = 340$ m/s was used. The combination of temperature and wind influence was computed as the sum of variances assuming statistical independence and a roughly Gaussian distribution.

The uncertainty of the modeling result can be derived from the measurement uncertainties using standard uncertainty propagation methods [144], see Appendix A.4.3.

---

[3]In order to show trends in the model data, here and in the next Table 4.2 it was refrained from rounding results to the significant number of decimals according to the estimated uncertainty.

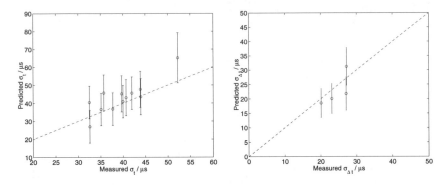

(a) Standard deviation of propagation time along a path.

(b) Standard deviation of propagation time difference along two paths.

Figure 4.24.: Predicted results relative to measured data, using wind velocity only ($\circ$), exact match (- -).

Since the contribution of the temperature fluctuations is small, only wind velocity data have been considered. A maximum measurement error of $\delta\tau = 1$ m was assumed for the correlation length, of $\delta\sigma_v = 0.1$ m/s for the standard deviation of the wind velocity, of $\delta s = 0.1$ m for the path length, and of $\delta c_0 = 3$ m/s for the speed of sound. However, the resulting uncertainty estimate is fairly insensitive with respect to the exact value of each uncertainty. The uncertainty of the measurement of the standard deviation of the propagation time was less than 1 μs and can be neglected.

Considering the results in Table 4.1 and Figure 4.24a a number of points can be noticed:

- The predicted standard deviation correlates with the measurement according to individual measurement setups, that is, measured and predicted values are approximately proportional (ideally their ratio would be 1).

- The influence of temperature fluctuations on the overall result is negligible. In fact, their standard deviation is roughly 10% of the standard deviation of the wind velocity, so it will contribute less than 0.5% to the combined standard deviation.

- All measurements fit the model results within the given range of error. The general trend represented by the data points is significant.

It can be concluded that these results demonstrate well the dependency of the standard deviation of the arrival time on the propagation distance as well as on the statistical properties of the wind field. To summarize one may state that although the number of measurement points is relatively small, eq. (4.16) is well reproduced by the measurement results, both with respect to the order of magnitude of the results as well as with respect to the dependency on two out of three primary parameters, namely propagation distance s and standard deviation of wind velocity σ_v.

Discussion of Errors

The results suggest that the derived correlation lengths of ca. 5 m for the air flow and ca. 2 m for the temperature are approximately correct. Unfortunately, the detailed dependence of the arrival time variance on the correlation length could not be tracked continuously with the available measuring equipment. For the long-distance measurement only two anemometers were used to capture the statistical data over a path of 15 m. Considering correlation lengths of 5 m, this setup can only deliver approximate coverage.

It should be mentioned that assuming an exponential relationship for the spatial correlation function is plausible but somewhat arbitrary. The scatter of the experimental values does not permit any rigorous conclusion regarding the detailed shape of the correlation functions. Gaussian or similar relationships are commonly used as well [65]. Within the uncertainty of the measurements presented here, the related differences are probably irrelevant. This aspect will be discussed next, in Section 4.4.2.

With respect to the measured standard deviation of the wind velocity, the following comments should be added:

- The variation of the standard deviation of the wind velocity along a single propagation path can be seen in legs 7 to 10 and 24 (Table A.2). Within the same measurement leg, the measurements of the standard deviation vary over a range of about ±0.07 m/s. Any local measurement that is considered representative for the propagation path is not more accurate than that.

- Another factor of influence seems to be the absolute accuracy of the measurements. Consider legs 13, 14, and 29 where the anemometers were located as close to each other as physically possible. Still the standard deviation shows a difference of up to ±0.03 m/s. Systematic offsets may be partially due to the different types of device being used. But they can also be caused by local, small eddies in the direct vicinity of the anemometer and its mounting.

- The measurement uncertainty of the anemometers throughout a leg can be neglected. According to the manufacturers a single measurement of the wind speed is accurate within a few percent. Given a set of several thousand measurements, the uncertainty of the standard deviation over that set will be less than 1% or 0.01 m/s.

- It is noteworthy that even under macroscopically identical conditions, that is, under approximately equal values for wind direction and velocity, such as in legs 9 and 10, or legs 7, 11, and 14, the structure of turbulence can differ. In some cases, the respective longitudinal and transversal standard deviations deviate to a degree. However, this effect is not important for the travel time calculations here since the standard deviations were measured explicitly and were not derived from the macroscopic parameters of the wind field.

These findings combined suggest that an uncertainty of about 0.1 m/s has to be assumed for the measured standard deviation of the wind velocity, σ_v.

Other errors worth mentioning include that an ergodicity assumption was made, meaning that conclusions are drawn regarding the spatial field of fluctuations from time-based measurements. In the sense of statistical physics temporal and spatial averages have been equated which seems justified in this model. Also, in both the presented model as well as the measurements any dispersion relations have been neglected. It has been assumed that under the condition of geometrical acoustics, the main effect of the environmental fluctuations on the propagation time will be independent of the wavelength.

The measurement data were also limited in the way that measurements were performed primarily in the horizontal domain and at relatively low wind speeds. Nonetheless these conditions are typical for the majority of practical sound system setups.

With respect to the measured values one may state that these are in good agreement with published data on the statistical properties of wind and temperature fluctuations in the atmosphere under comparable conditions [61], [62], [131], [145], [146], [147].

4.4.2. Propagation Time Difference

In legs 21 to 24 the propagation time difference was measured by recording the signal from different loudspeakers simultaneously. The theoretical model, eq. (4.29), that was developed earlier can be used to predict the standard deviation of the travel time difference. It requires as input data the correlation length and the standard deviation of environmental parameters which were also measured.

Table 4.2 shows the path length s for each signal and the base angle γ of the two propagation paths. It also reports the measured standard deviation $\sigma_{\Delta t}$ as well as the computed values, both for the numerical integral and its approximation. For the speed of sound a fixed value of $c_0 = 340$ m/s was assumed. In this validation the small effect of temperature fluctuations was neglected and only wind influence was considered. The correlation length was assumed to be $\tau = 5$ m and for the standard deviation of the wind velocity σ_v the average value was taken of the two longitudinal measurements made. The same data are visualized in Figure 4.24b.

The measurement data and the modeled values show good agreement. On the one hand, both data sets are clearly of the same order of magnitude and deviate from each other only in the range of 20% and within the given error range. On the other hand, the values show the same tendency. They increase for a larger base angle and for longer traveling distances. However, one should also state that this set of only four measurement series is limited in its ability of supporting the theoretical model. More measurements using a broader range of values for the input parameters would be helpful for the validation.

4. Coherence of Radiated Sound

Leg	Lsp1	Lsp2	Angle	Measured	Modeled Exact	Modeled Approx.	Modeled Uncertainty
	s_1/m	s_2/m	$\gamma/°$	$\sigma_{\Delta t}/\mu s$	$\sigma_{\Delta t}/\mu s$	$\sigma_{\Delta t}/\mu s$	$\delta\sigma_{\Delta t}/\mu s$
21	8.00	9.80	13	20.1	18.5	16.3	5
22	8.00	9.80	13	23.1	20.1	17.6	5
23	8.00	8.90	25	27.1	21.8	19.6	6
24	8.75	14.35	0	27.2	31.3	31.3	7

Table 4.2.: Measured and modeled standard deviation of propagation time difference for various measurement setups. Velocity part only. See also Fig. 4.24b.

The uncertainty estimates were determined in the same manner as for the propagation time (see also Appendix A.4.3). The uncertainty of the base angle was assumed to be $\delta\gamma = 1°$. Because the uncertainty estimates were derived based on the linear approximation model, a correction factor was included that accounts roughly for the offset from the exact, numerically computed result for the respective parameter combinations.

Discussion of Errors

Besides the limitations discussed before it should be remarked that the range of values for the propagation distance was limited. Due to practical circumstances a maximum distance of about 15 m could not be exceeded during the measurements.

In contrast, the range of angles covered is fairly representative. For much wider angles the measurements will be dominated by the uncorrelated part in the overall standard deviation. If the correlation between propagation on different paths has to be evaluated, a good portion of the two paths should be overlapping within the correlation length. At an angle of 25° the loudspeakers were about 3.7 m apart which is in the order of the correlation length.

Comparison with Existing Work

The results of the theoretical model for the standard deviation of the travel time difference given by equations (4.16) and (4.29) can be compared with other published results. In order to clearly distinguish the model for the field of fluctuations presented in this work from other approaches, it will be called *exponential decay* model in the following. It should be noted that several published studies of this topic [61], [62] are based on or related to the analysis by Ostashev [65] and reproduce his results only in a relatively incomplete form.

In the context of that theory, the *Gaussian* and the *von Karman* spectrum models for the field of fluctuations can be considered as commonly used and as practically relevant for a comparison. But when comparing with these models it has to be emphasized that most results are given for very large distances under the assumption that the Fraunhofer condition $D = 2s\lambda/(\pi\tau^2) \gg 1$ is valid (see also Section 3.1.1). This is not necessarily true in the case of the exponential decay model presented here. Even though it was

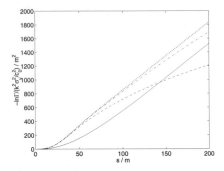

Figure 4.25.: Coherence factor using Gaussian spectrum model, 315 Hz (- -), 1000 Hz (-.-), 3150 Hz ($\cdots$), and exponential decay model ($-$), $\gamma = 10°$, $\tau = 5$ m, wind velocity only.

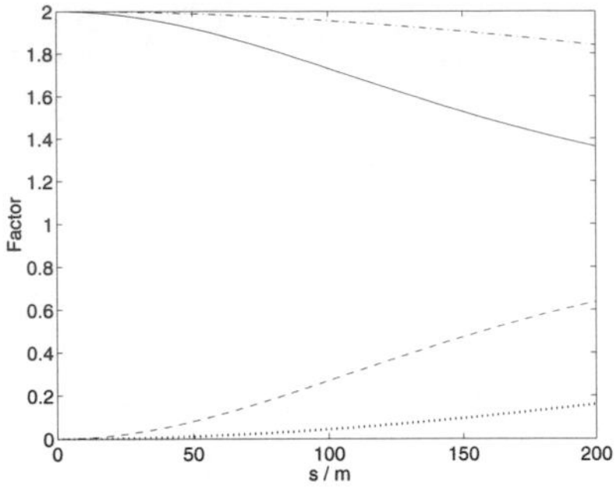

Figure 4.26.: Relative strength of fluctuations according to Gaussian spectrum model, phase 315 Hz ($-$) and 1000 Hz (-.-), amplitude 315 Hz (- -) and 1000 Hz ($\cdots$), $\tau = 5$ m, wind velocity only.

derived under the condition of geometrical acoustics, $\lambda \ll \tau$, and its results are also valid in the far field, $s \gg \tau$, this may or may not imply the Fraunhofer condition $s/\tau \gg \tau/\lambda$. On the contrary, here the primary interest is in D values of the order of 1.

Furthermore, care must be taken when comparing quantities computed from different statistical models. Structure parameters that are used to define the field of fluctuations, such as the characteristic dimension of turbulence τ, are not directly transferable between models. In the considerations here the calibration method of Ostashev has been followed which yields $\tau = \frac{\sqrt{\pi}}{2} l_G$, where l_G is the characteristic scale of turbulence for the Gaussian spectrum.

Figure 4.25 shows a comparison of the normalized variance for the Gaussian spectrum model with the corresponding data for the exponential decay model. The quantity shown is the natural logarithm of the phase part of Ostashev's coherence factor, $-\ln(\Gamma) = \langle \phi^2 \rangle - B_\phi$. This equates to $\omega^2(\sigma_t^2 - \alpha^2 \sigma_T^2 I)$ in the terminology of the exponential decay model, eq. 4.29, under the conditions of $s = s_1 = s_2$ and $\sigma_T = \sigma_v$. For better display the resulting value was normalized to $k^2 \sigma_v^2 / c_0^2$.

The Gaussian curves were computed for three different wavelengths λ. They are based on the relationships given by Ostashev for the spherical wave form and include only the part describing the phase fluctuations. Temperature fluctuations have been left out and the comparison has been restricted to wind velocity only, since it shows the largest deviations between the models. This is partly due to the fact that the Gaussian spectrum model accounts for anisotropy in the field of fluctuations relative to the traveling direction of the sound wave.

The functions in Figure 4.25 show a similar behavior with respect to distance s for an exemplary set of parameters. For very large distances relative to the wavelength, $D \gg 1$, the Gaussian model approaches an asymptotic course that differs from the slope of the exponential decay model by a factor of $1/2$. However, in the range of distances and frequencies that are important for sound reinforcement applications, $D \approx 1$, the exponential decay model is clearly within the spread of the Gaussian set. Unfortunately, even in [65] the coherence factor and related quantities for the von Karman spectrum are given only for the asymptotic case $D \gg 1$. An illustration of these functions is provided in Appendix A.4.4. Nonetheless it should be remarked that the von Karman spectrum model is very similar to the exponential decay model, since it also exhibits an exponential decay over distance. Based on the similarity of the results of the exponential decay model with the results of the Gaussian spectrum one would expect an even closer match for the von Karman spectrum.

With respect to the parameters of Figure 4.25, more results for different angles γ as well as for temperature fluctuations are given in Appendix A.4.4. Notice that the relationships derived in [65] are subject to the condition of small angles $\theta \ll 1$, where $\theta = \gamma/2$, and are therefore limited to $\gamma \lesssim 15°$.

It was stated above that this comparison is restricted to phase fluctuations. Amplitude fluctuations have been neglected in the derivation of the exponential decay model and they have been excluded from the comparative computations based on the Gaussian spectrum model as well. Figure 4.26 shows a comparison of the variance of the phase fluctuations $\langle \phi^2 \rangle$ relative to the variance of the log-amplitude fluctuations $\langle \chi^2 \rangle$ according to [65]. In the terminology of this publication the quantities $\langle \phi^2 \rangle \sim 1 + N(D)$ and $\langle \chi^2 \rangle \sim 1 - N(D)$ are depicted. First of all, one can recognize that for typical distances and frequencies of interest the amplitude fluctuations are significantly smaller than the phase fluctuations. Interestingly, there are also investigations [61] that indicate that over larger distances amplitude fluctuations can be neglected relative to phase fluctuations. Therefore at this time it remains future work to examine the role of amplitude fluctuations in practice.

Comparing the weight of temperature fluctuations relative to wind velocity fluctuations one may state that all three models show a quantitatively similar behavior. Under the above normalization the travel time variance caused by temperature fluctuations is smaller by a factor of 4. The respective values for σ_θ^2/T_0^2 and σ_v^2/c_0^2 are different by an additional factor of about 20 for typical environmental parameters. Overall the travel time variance attributed to temperature fluctuations relative to the wind velocity fluctuations is about 1%, or with respect to the standard deviation of the travel time 10%. Also, all models assume that wind fluctuations and temperature fluctuations are independent of each other.

In all comparisons above, analytic investigations were limited to the asymptotic behavior of the coherence factor under various conditions. Due to the complexity of the relationships that define the coherence factor and its related quantities in the considered models, the exact course of the functions has been compared only numerically for a number of cases that cover the parameter range of interest.

To sum it up one can say that the results delivered by the exponential decay model are quantitatively in the same range as those of other published models. They also show the same qualitative behavior as far as these models can be evaluated. The range of validity of the exponential decay model overlaps but is not identical with that of the cited models.

4.5. Practical Applications

Two Remote Sources

There are several practical consequences of the findings presented before that are worth being discussed. The relationships defined by equations (4.45) and (4.51) show that for any given scatter of arrival times of propagating sound due to air fluctuations there is a loss of mutual coherence among identical signals emitted from multiple acoustic sources. Especially for the case of two signals that are in-phase if no environmental effects are present, one can compute the statistical level reduction due to air flow fluctuations, for example. Inserting equation (4.45) into (4.51) and thereby considering $\sigma_\xi^2 = \sigma_{\xi_1}^2 + \sigma_{\xi_2}^2$ under the condition of uncorrelated propagation paths yields the following relationship for the average power:

$$\langle P \rangle = 2|A|^2 \Big[1 + \exp(-\omega^2 \sigma_t^2) \Big]. \tag{4.56}$$

The mean power $\langle P \rangle$ can be normalized to the fully incoherent power sum, $P_{Ran} = 2|A|^2$, and combined with the travel time variance σ_t^2 in eq. (4.16):

$$\frac{\langle P \rangle}{P_{Ran}} = 1 + \exp\left\{ -2\omega^2 \alpha^2 \sigma_T^2 \tau \Big[s + \tau \big(e^{\frac{s}{\tau}} - 1 \big) \Big] \right\} \tag{4.57}$$

where s is the propagation distance. The exponential term represents the transition between the random-phase state, where it vanishes, and the fully coherent state, where it reaches unity. As a generalization of this interesting quantity one can define the relative coherence C as

$$C = \frac{\langle P \rangle - P_{Ran}}{N(N-1)|A|^2}, \tag{4.58}$$

where N is the number of sources and $0 \leq C \leq 100\%$. Figures 4.27a and 4.27b illustrate this relationship in two different ways, namely as the relative coherence C in the signal duplet, and as a reduction of the relative sum level of the two sources given by $L = 10 \log(\langle P \rangle / |A|^2)$.

Data were computed for typical wind parameters; standard deviation of longitudinal wind velocity was $\sigma_v = 0.64$ m/s and correlation length was $\tau = 5$ m, like in the measurement data shown earlier. Temperature effects were not included.

Evidently, coherence is increasingly lost for higher frequencies and longer propagation distances. This quantitative behavior was so far only qualitatively familiar among audio practitioners. Of course, there is also the inverse effect to the level reduction for in-phase signals. If signals are phase-shifted by 180° the loss of coherence will lead to a level increase. In the limit of random phase any signal cancelations, like comb filters, will be smoothed accordingly. Obviously, both effects are not very desirable for modern line array systems whose main intent is to focus sound energy at the audience and keep it away from reflective surfaces at the same time.

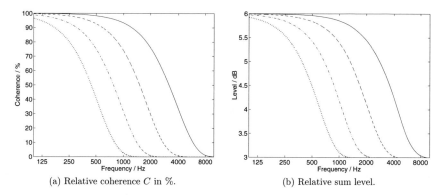

(a) Relative coherence C in %.   (b) Relative sum level.

Figure 4.27.: Reduction of the combined signal from two remote sources, as a function of frequency and propagation distance, 300 m ($\cdots$), 100 m (-.-), 30 m (- -), 10 m (—). In this typical example, the standard deviation of wind velocity is $\sigma = 0.64$ m/s and the correlation length τ is 5 m.

For the given wind parameters and at a receive distance of 100 m, coherence is basically lost for frequencies above 2 kHz. Similarly, at distances larger than 10 m, the loss of coherence becomes notable in the audible spectrum.

As a rule of thumb for sources spaced far apart from each other one can define the *critical frequency* at which the coherence is reduced to 50%:

$$f_{Crit} \approx \frac{10}{\sqrt{s/m}} \text{kHz} \tag{4.59}$$

where s is the receive distance and must be at least 10 m. Here a spatial correlation length of wind velocity of 5 m and a standard deviation of the order of 0.5 m/s (light breeze) are assumed.

On the same basis one can derive the *critical distance* beyond which the coherence drops below 50%:

$$s_{Crit} \approx \frac{100}{(f/\text{kHz})^2}\text{m} \tag{4.60}$$

where f is the considered frequency limit in kHz. Like above, this relationship is a rough approximation for light wind and receive distances larger than 10 m.

Considerations in this section were limited to two sources of equal strength at the receiver and uncorrelated propagation paths. Those results can be generalized to more than two different sources in a straight-forward manner. It should also be emphasized that the travel time equation of geometrical acoustics (4.1) only holds as long as the inhomogeneities are small deviations from the equilibrium and spatially larger than a wavelength. Further, it has been assumed in this discussion that amplitude fluctuations are negligible.

Line Array Elements

Similar to Fig. 4.27a, Fig. 4.28a shows a coherence curve over frequency but here results are depicted for an array of 8 identical source elements and of a height of 8 m. The curves are computed using eq. (4.55) and it is assumed that the receiver is located on-axis of the array and that all path lengths are approximately equal. Unlike Fig. 4.27a, correlations between propagation paths are accounted for in this setup. Between adjacent paths, the correlation of air fluctuations within a finite radius enhances the array's overall coherence and extends it toward higher frequencies or longer distances, respectively.

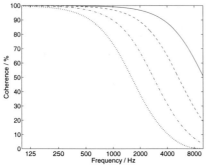 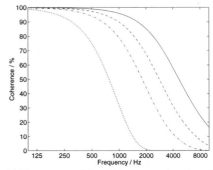

(a) Line array of 8 m, as a function of frequency and propagation distance, 300 m ($\cdots$), 100 m (-.-), 30 m (- -), 10 m (—).

(b) Line array at 100 m distance, as a function of frequency and array size, 16 m (-.-), 8 m (- -), 4 m (—), uncorrelated propagation paths ($\cdots$).

Figure 4.28.: Relative coherence C in %, eq. (4.58). In this typical example, the standard deviation of wind velocity is $\sigma = 0.64$ m/s and the correlation length τ is 5 m. The number of array elements is $N = 8$.

Figure 4.28b illustrates the dependence of the loss of coherence on the size of the array relative to the correlation length (here, $\tau = 5$ m). The longer the array, the stronger the influence of the environmental fluctuations and the weaker the interference effects between the most distant array elements. As a reference the corresponding curve for uncorrelated propagation paths is also shown, eq. (4.57). By definition this curve represents the lower limit for the coherence of extended sources, namely a very small correlation length relative to the array length.

In practice, the results shown in Fig. 4.28 mean that large line arrays of about 10 m and taller suffer a significant performance loss for the frequency range above 2 kHz at distances of 100 m and beyond. For arrays of 10 or more elements, a coherence of 50% corresponds to a loss of about 3 dB and 25% to a loss of 6 dB eq. in level, eq. (4.58).

The above results appear robust with respect to the number of simulated array elements as long as that number is of the order of 10 or greater. For an increasing amount of sources the convergence to the same, asymptotic results is fairly rapid. This indicates that the sum in eq. (4.55) can be generalized to an integral. An in-depth analysis of this aspect is beyond the scope of this work, but some related details can be found in [135].

4.6. Summary and Open Problems

In this chapter a closed, analytical expression, eq. (4.16), was derived that defines the standard deviation of the propagation time of sound in an inhomogeneous medium as a function of the statistical properties of the wind vector and of the air temperature. Also, an integral expression was obtained for the standard deviation of the propagation time difference between two sound waves traveling along separate paths that are influenced by mutually correlated fluctuations, eq. (4.29). Using linear approximations a simplified analytical expression could be derived that is fairly accurate over the largest part of the space of definition and represents a lower bound at the same time. For both main results the method of statistical ensemble averages was used which was also employed to take a second step, namely to quantify the average effect of the fluctuations in the propagation delay on the coherent superposition of the signal of two or more sources.

After that the analysis of numerous results from outdoor measurements of environmental conditions as well as of the propagation of sound was presented. In addition to providing basic statistical quantities, the data also allowed deriving typical correlation times and lengths for the wind velocity. Comparison of the wind data with measurements of the propagation delay of sound along the same axis showed a strong correlation. In contrast, the effect of measured temperature variations on the arrival time was negligibly small.

Subsequently measured data for the standard deviation of the propagation time along a single path as well as for the standard deviation of the travel time difference of signals on two different paths were compared against theoretical predictions based on the statistical properties of the field of wind fluctuations. It was shown that the analytical results match well with the experimental data within the limitations of the theoretical derivation and the measurement. Additionally, comparison of the results of the theoretical model presented in this work with published results based on other models showed good agreement.

It can be concluded that the mathematical relationships derived in this work can be used to estimate the detrimental effects of wind and temperature fluctuations on the coherence of acoustic signals. The presented rules of thumb, eqs. (4.57) - (4.60), can be used as practical estimates of the coherence loss. The detailed results are particularly interesting for acoustic modeling software packages, such as [92] and [148], in order to facilitate a more accurate prediction of the performance of larger, distributed sound systems. Although the presented formulae are only approximations, they establish the first, practically usable means to replace the commonly used on/off switch for signal coherence by a novel function that depends on receive distance, loudspeaker separation, and frequency.

Also, some of the limitations of the analysis presented here should be summarized:

- In this chapter the focus was on the effect of small fluctuations in the propagation medium in the high-frequency limit of the propagation of sound. In this model wave-based and non-local propagation characteristics cannot be taken into account.

- The conducted experiments cover only a part of the parameter range encountered in practical use cases. A number of theoretical assumptions were made using plausible arguments, such as regarding the correlation functions, but they would require substantially more experimental data in order to be thoroughly validated.

- Also, no measurements were performed of the average sound pressure level of combined sources. This would be useful in order to directly demonstrate the practical value of the model described by eqs. (4.51) and (4.55).

- The discussion in this study was concerned with levels averaged over time periods that are significantly longer than the duration of the fluctuations. Of course, looking at the sequence of several points of time the instantaneous transmission channel will change with the rate of the fluctuations (ca. 5-10 s). Accordingly the instantaneous impulse response and the derived acoustic quantities will drift in the same manner. In comparison to that, a long-term average of these data may not be able to adequately represent all subjective aspects of the transmission quality since human hearing averages over much shorter periods of time.

- All of the considerations were restricted to properties of the transmission paths. However, in practice also the temporal shape of the transmitted signals will play a role in the perceived influence of the loss of coherence. When using signals with very diverse content and (in comparison) quickly changing amplitudes, fluctuations of the travel time may be less notable than for comparably long signals with a high self-similarity where the transient changes of the transmission path can be recognized. One may also recognize that very short, impact-like or percussive signals sound very focused at one point of time and smeared out at a different point of time due to the fluctuations in relative arrival times.

5. Conclusions and Outlook

5.1. Summary and Conclusions

In this work a wide range of topics related to the modeling of sound sources in the context of acoustic simulation was covered. Although many of the discussions were based on theoretical arguments and mathematical derivations, in part even very detailed ones, this thesis is aimed at the development of a thorough and comprehensive framework for the practical modeling of the radiation of modern sound systems. This includes the mathematical treatment of the underlying physical processes and circumstances as well as the practice-oriented engineering of measurement conditions, resolution requirements, and data representation.

The key results of this work relate to three different but connected fields of application: the acoustic simulation of sound radiated by small sound sources, by large sound sources, and by combinations of sound sources.

Small sound sources, such as individual loudspeakers, are used in their far field, that is, at distances much larger than the spatial extent of the acoustic source itself. It was shown that for the combination of such sound sources in the model of a multiway loudspeaker or of an array of loudspeakers, for the accuracy of simulation results it is crucial to include directional phase information along with magnitude data. The description of sound transducers by using complex data also turns the traditional search for the so-called acoustic center unnecessary and irrelevant. A set of conditions for the measurement of such complex directional data was derived as well as proposed methods for processing these data. This led to the formal definition of the Complex Directivity Point Source (CDPS) model for sound sources. The Generic Loudspeaker Library (GLL) description language was presented as a practical means to implement the CDPS model. The GLL was defined as an object-oriented, expandable data format with the objective to provide the degree of flexibility and configurability required for the representation of concurrent loudspeaker systems in the modeling domain.

Large sound sources, such as concert sound line arrays, are mostly used into their near field. A line array cannot be treated like a point source in typical modeling scenarios. For this application, it was demonstrated by means of the CDPS decomposition method that theoretically any approximate, finite line source can be subdivided into smaller elements for which individually the far-field assumption can be applied successfully. This permits conducting directional measurements on the elemental level and modeling line arrays based on a CDPS model for each cabinet or transducer. The CDPS decomposition also provides a way to derive practical requirements for the spatial and spectral data resolution of the model. Accounting for that will yield highly accurate prediction results, given that the directional measurements of a line array element include the acoustic influence of the circumjacent physical structure, e.g. by adding two inactive cabinets to the setup which are arranged as top and bottom neighbors. With respect to arrays of sound sources it is also of much practical interest to evaluate the effect of variation among supposedly identical elements on the overall system performance. Analysis and simulation showed that the relative error in the overall received sound pressure will generally be reduced by a factor proportional to the square root of the number of elements. This finding was underpinned by several measurement data sets from the production lines of loudspeaker manufacturers. The latter is particularly remarkable because such data are often considered as politically sensitive and are published very rarely.

Finally, a comparably simple model was introduced for simulating the effect of small environmental fluctuations, such as air flow or temperature changes, on the coherence of the signals of several sound sources superimposed at the receive location. The analysis was conducted on the basis of geometrical acoustics and resulted in mathematical expressions for the variance of the travel time along a propagation path and for the variance of the travel time difference between two different propagation paths. Within the given limitations the theoretical results matched well with measured data as well as with other published results. Based on this statistical spreading of propagation times one can draw conclusions about the received average sound pressure level of a group of loudspeakers as a function of receive distance, frequency, and the statistical properties of the environmental parameters. This more sophisticated approach allows replacing the conventionally used models of either power summation or complex summation of pressure amplitudes applied to the entire frequency range and coverage area.

Since the introduction of the CDPS model and the GLL data format in 2006 this concept has found widespread acceptance in the pro-audio industry. Nowadays most established loudspeaker companies and laboratories generate loudspeaker modeling data in the GLL format and account for the required measurement conditions. Parts of this work related to loudspeaker data representation were implemented in the AES56 standard of the Audio Engineering Society. A publication introducing the CDPS decomposition model as detailed in Chapter 3 was awarded with the renowned biennial AES Publications Award in 2010. These distinguished events illustrate that the research work presented here was generally recognized and well received by audio professionals and has already gained broad practical significance on the international level.

5.2. Future Work and Outlook

This work summarizes the results of research work from 2005 to 2012 and still on-going. Future developments in DSP processing, computer hardware and sound reproduction technology will inevitably lead to corresponding developments in the modeling domain.

In the previous chapters it has been indicated already that most electro-acoustic and room acoustic modeling software are lacking even the most basic means to predict the uncertainty of the calculation results. While establishing uncertainty estimates may be quite difficult for complicated ray-tracing computations it seems rather straight-forward for models of the direct sound field. In this respect, uncertainties with respect to the input data, such as measurement uncertainty and production tolerance, should be stored along with the loudspeaker data and processed in the calculation runs in order to provide the end user with a clear picture of the reliability of the simulation results.

A completely new, real-world development direction is given by combining high-resolution simulation capabilities with advanced DSP processing, e.g. FIR filtering. Assuming accurate loudspeaker data as input, optimum filter settings for each loudspeaker or DSP channel can be computed on the basis of the geometrical model of the venue [40], [45], [126]. This approach can improve the performance of line arrays and loudspeaker clusters dramatically and may significantly reduce the time required for on-site tuning and commissioning. As a prerequisite, simulation models have to be extended from including loudspeaker data and geometry information to virtual representations of the DSP and amplifier network.

Another point worth mentioning is the modeling of sound sources in the low-frequency domain. It was indicated in Chapters 1 and 3 how methods like BEM, FEM, or FDTD can be used to simulate wave-based effects in rooms or loudspeaker arrangements. It will be interesting to see whether these methods can be extended toward higher frequencies and supported by improved definitions of the boundary conditions. A particularly interesting aspect is, for example, the interaction of subwoofer loudspeakers with each other and with the room.

These few items already show that the topic of modeling the radiation of loudspeakers and sound systems will not be closed at any foreseeable time soon. Hopefully, the development of new technology solutions will go hand in hand with better theoretical models and advancing computer simulation capabilities.

5.3. Acknowledgments

On the basis of a number of published articles this work was compiled over the course of the years 2011 and 2012. It is largely related to research that I have conducted in parallel to software development projects at the companies AFMG and SDA.

I am very grateful to Prof. Michael Vorländer for supporting this work, for providing a number of essential hints, and for making it possible to develop and complete this thesis as an external researcher. I would like to thank Prof. Dirk Heberling, the second reviewer of this work, for his interest and his advice, as well as my examiners Prof. Andrei Vescan and Prof. Wilfried Mokwa.

I also wish to thank the co-authors of the publications related to this work: Prof. Wolfgang Ahnert, my mentor, partner and friend, provided many practice-related comments and valuable ideas. Ambrose Thompson from Martin Audio contributed numerous thoughts, line array measurements as well as figures to [50] some of which were reproduced in Chapter 3.3. Our year-long, intense collaboration was eventually distinguished by the above-mentioned AES award. Charlie Hughes and Bruce Olson made various loudspeaker measurements and gave me more insight in the loudspeaker design and optimization process [100], as described in Chapter 2.4.3. Dr. Steffen Bock implemented a significant part of the early versions of the GLL format in usable source code [48].

This work would have been impossible without the measurement data and partly the support work provided by many different people and companies. I am much obliged to Ralph Heinz and Brandon Mosst from Renkus-Heinz, to Bill Gelow and Mike Kasten from Electro-Voice, to José Ramón Menzinger and Martin Kling from Kling & Freitag, to Jason Baird and Phil Knight from Martin Audio, to Ron Sauro from NWAA Labs, and to Tom Back from Alcons Audio.

I am also grateful for motivating discussions with Harro Heinz from Renkus-Heinz, Prof. Anselm Goertz and Dr. Michael Makarski from Four Audio, Tomlinson Holman, Hiroshi Kubota from TOA, Dominic Harter from Turbosound, Jim Brown, Joe Brusi, Mario Di Cola, and Daniele Ponteggia.

Finally, I would like to express my deepest thanks to my family, especially to my wife, Dr. Angela Feistel, for her support and understanding, and to my father, Dr. Rainer Feistel, for many inspiring discussions over the last years and numerous helpful hints in the final phase of this work.

A. Appendix

A.1. Acronyms and Abbreviations

AES	Audio Engineering Society
BEM	Boundary Element Method
BIR	Binaural Impulse Response
CDPS	Complex Directivity Point Source
DI	Directivity Index
DLL	Dynamic Link Library
DSP	Digital Signal Processor
EQ	Equalizer
FDTD	Finite-Difference Time-Domain
FEM	Finite Element Method
FFT	Fast Fourier Transform
FIR	Finite Impulse Response
GLL	Generic Loudspeaker Library
HF	High-Frequency
HRTF	Head-Related Transfer Function
LF	Low-Frequency
MF	Mid-Frequency
PC	Personal Computer
POR	Point of Rotation
SPL	Sound Pressure Level
XTC	Cross-Talk Cancelation

Table A.1.: Acronyms and abbreviations used in this work.

A.2. GLL Project Example

Fig. A.1 shows a typical section of the project text file of a line array GLL. The format is object-oriented and based on a key-value-pair scheme like XML.

```
"GLL"
"Format", "3D"
"FormatVersion", "1.02"

"System", "Sphere Line", "SL", "LA"
"SystemVersion", 1.10
"Company", "AFMG"

"InfoText", "EASE SpeakerLab Example - This is an imaginary array. Similarities to existing
systems are not intended.\r\n\r\nUses omnidirectional sources."
"CopyrightText", "© 2007-2010 AFMG"
"SupportText", "See manual for support and rigging instructions"
"WebsiteText", "www.afmg.eu"
"EmailText", "info@afmg.eu"

"Logo1", "..\afmg.png"
"Logo2", ""
"BackgroundColor", 0

"BoxTypes", 4
    "BoxType", "Big", "keyBoxBig"
        "Horizontal Opening Angle", 100
        "Vertical Opening Angle", 40
        "Sources", 3
            "Source", "MB", "1", 0.20, 0.00, -0.20, 0.00, 0.00, 0.00, "12 inch"
            "Source", "Horn", "2", 0.00, 0.00, -0.20, 0.00, 0.00, 0.00, "60x40"
            "Source", "MB", "3", -0.20, 0.00, -0.20, 0.00, 0.00, 0.00, "12 inch"
        "Input Configurations", 1
            "Input Configuration", "Passive Mode", "PassiveMode"
                "Inputs", 1
                    "Input", "Input"
                        "Rated Impedance", 8.00
                        "Links", 3
                            "Link", "1", ""
                            "Link", "2", ""
                            "Link", "3", ""
        "GeometryFiles", 1
            "GeometryFile", ".\SymmFed\CaseBigBox.fed"
        "ReferencePoint", 0.00, 0.30, -0.20
        "NextPivot", 0.00, 0.00, -0.40
        "CenterOfMass", 0.00, 0.30, -0.20
        "Weight", 25.00

    "BoxType", "Small", "keyBoxSmall"
```

Figure A.1.: Part of a GLL project file.

A.3. Further Statistical Data for Line Arrays

A.3.1. Omniline Statistical Data

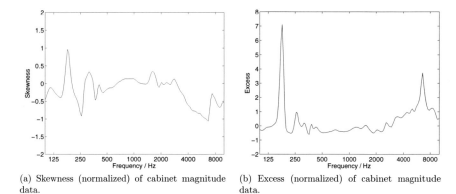

(a) Skewness (normalized) of cabinet magnitude data.

(b) Excess (normalized) of cabinet magnitude data.

Figure A.2.: Further statistical properties of Omniline cabinet samples.

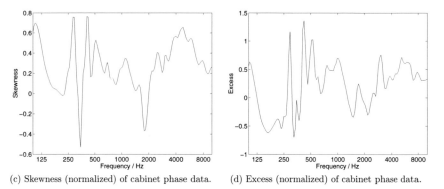

(c) Skewness (normalized) of cabinet phase data.

(d) Excess (normalized) of cabinet phase data.

Figure A.2.: *Continued.*

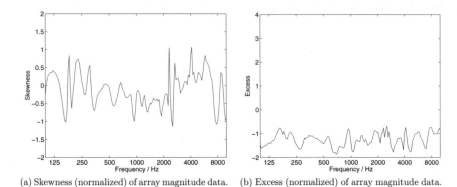

(a) Skewness (normalized) of array magnitude data.   (b) Excess (normalized) of array magnitude data.

Figure A.3.: Further statistical properties of Omniline array samples.

A.3.2. Iconyx IC-8 Statistical Data

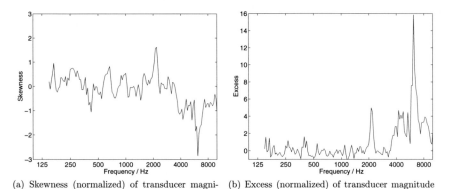

(a) Skewness (normalized) of transducer magnitude data.   (b) Excess (normalized) of transducer magnitude data.

Figure A.4.: Further statistical properties of IC-8 transducer samples.

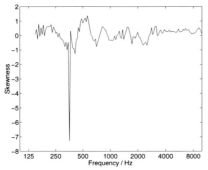 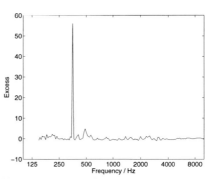

(c) Skewness (normalized) of transducer phase data.   (d) Excess (normalized) of transducer phase data.

Figure A.4.: *Continued.*

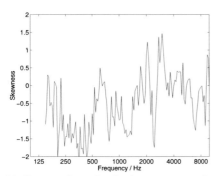 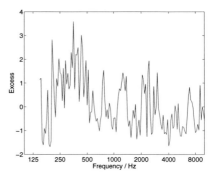

(a) Skewness (normalized) of array magnitude data.   (b) Excess (normalized) of array magnitude data.

Figure A.5.: Further statistical properties of Iconyx IC-8 array samples.

A.4. Coherence of Radiated Sound

A.4.1. Analytical Derivations

Approximation of Correlation Integrals

Case $s_2 \geq s_b(s_1) = s_1/v$:

$$I = \int_0^{s_1} ds \left[\int_0^{s_a(s)} ds' e^{-r^{(1)}/\tau} + \int_{s_a(s)}^{s_b(s)} ds' e^{-r^{(2)}/\tau} + \int_{s_b(s)}^{s_2} ds' e^{-r^{(3)}/\tau} \right]. \tag{A.1}$$

One calculates the inner integrals first:

$$J_1 = \int_0^{sv} ds' e^{-r^{(1)}/\tau} = \frac{\tau}{\cos \gamma} e^{-s/\tau} \left(e^{sv \cos \gamma/\tau} - 1 \right), \tag{A.2}$$

$$J_2 = \int_{sv}^{s/v} ds' e^{-r^{(2)}/\tau} = -\frac{\tau}{\sin(\gamma/2)} e^{-s \sin(\gamma/2)/\tau} \left(e^{-s/v \sin(\gamma/2)/\tau} - e^{-sv \sin(\gamma/2)/\tau} \right), \tag{A.3}$$

$$J_3 = \int_{s/v}^{s_2} ds' e^{-r^{(3)}/\tau} = -\tau e^{-s \cos \gamma/\tau} \left(e^{-s_2/\tau} - e^{-s/(v\tau)} \right). \tag{A.4}$$

The outer integrals are given by:

$$L_1 = \int_0^{s_1} ds J_1 = \frac{\tau^2}{\cos \gamma} \left[\frac{1}{v \cos \gamma - 1} \left(e^{s_1(v \cos \gamma - 1)/\tau} - 1 \right) + \left(e^{-s_1/\tau} - 1 \right) \right], \tag{A.5}$$

$$L_2 = \int_0^{s_1} ds J_2 = \frac{\tau^2}{\sin^2(\gamma/2)} \left[\frac{1}{1 + 1/v} \left(e^{-\sin(\gamma/2)(1+1/v)s_1/\tau} - 1 \right) \right.$$
$$\left. - \frac{1}{1+v} \left(e^{-\sin(\gamma/2)(1+v)s_1/\tau} - 1 \right) \right], \tag{A.6}$$

$$L_3 = \int_0^{s_1} ds J_3 = -\tau^2 \left[\frac{e^{-s_2/\tau}}{\cos \gamma} \left(e^{s_1 \cos \gamma/\tau} - 1 \right) - \frac{1}{\cos \gamma - 1/v} \left(e^{(\cos \gamma - 1/v)s_1/\tau} - 1 \right) \right]. \tag{A.7}$$

The sum of the outer integrals represents the final result:

$$I = L_1 + L_2 + L_3. \tag{A.8}$$

A. Appendix

Case $s_1/v = s_b(s_1) \geq s_2$:

$$I = \int_0^{s_2 v} ds \left[\int_0^{s_a(s)} ds' e^{-r^{(1)}/\tau} + \int_{s_a(s)}^{s_b(s)} ds' e^{-r^{(2)}/\tau} + \int_{s_b(s)}^{s_2} ds' e^{-r^{(3)}/\tau} \right]$$

$$+ \int_{s_2 v}^{s_1} ds \left[\int_0^{s_a(s)} ds' e^{-r^{(1)}/\tau} + \int_{s_a(s)}^{s_2} ds' e^{-r^{(2)}/\tau} \right]. \quad (A.9)$$

One calculates the remaining inner integrals first (see above for J_1, J_2 and J_3):

$$J_4 = J_1, \quad (A.10)$$

$$J_5 = \int_{sv}^{s_2} ds' e^{-r^{(2)}/\tau} = -\frac{\tau}{\sin(\gamma/2)} e^{-s \sin(\gamma/2)/\tau} \left(e^{-s_2 \sin(\gamma/2)/\tau} - e^{-sv \sin(\gamma/2)/\tau} \right). \quad (A.11)$$

The outer integrals are given by:

$$K_1 = \int_0^{s_2 v} ds\, J_1 + \int_{s_2 v}^{s_1} ds\, J_4 = \int_0^{s_1} ds\, J_1 = L_1, \quad (A.12)$$

$$K_2 = \int_0^{s_2 v} ds\, J_2 = \frac{\tau^2}{\sin^2(\gamma/2)} \left[\frac{1}{1+1/v} \left(e^{-\sin(\gamma/2)(1+1/v)s_2 v/\tau} - 1 \right) \right.$$

$$\left. - \frac{1}{1+v} \left(e^{-\sin(\gamma/2)(1+v)s_2 v/\tau} - 1 \right) \right], \quad (A.13)$$

$$K_3 = \int_0^{s_2 v} ds\, J_3 = -\tau^2 \left[\frac{e^{-s_2/\tau}}{\cos \gamma} \left(e^{s_2 v \cos \gamma/\tau} - 1 \right) - \frac{1}{\cos \gamma - 1/v} \left(e^{(\cos \gamma - 1/v)s_2 v/\tau} - 1 \right) \right], \quad (A.14)$$

$$K_4 = \int_{s_2 v}^{s_1} ds\, J_5 = \frac{\tau^2}{\sin^2(\gamma/2)} \left[e^{-s_2 \sin(\gamma/2)/\tau} \left(e^{-s_1 \sin(\gamma/2)/\tau} - e^{-s_2 v \sin(\gamma/2)/\tau} \right) \right.$$

$$\left. - \frac{1}{1+v} \left(e^{-\sin(\gamma/2)(1+v)s_1/\tau} - e^{-\sin(\gamma/2)(1+v)s_2 v/\tau} \right) \right]. \quad (A.15)$$

The sum of the outer integrals establishes the final result:

$$I = K_1 + K_2 + K_3 + K_4. \quad (A.16)$$

A. Appendix

Upper Bounds

Approximation (1) is an upper bound because the expression in the exponent is a lower bound:

$$r_{(1)} = s - s' \cos \gamma, \tag{A.17}$$

$$r_{(1)}^2 = s^2 + s'^2 \cos^2 \gamma - 2ss' \cos \gamma \leq s^2 + s'^2 - 2ss' \cos \gamma = r^2. \tag{A.18}$$

The same is true for approximation (3) since one can simply exchange s and s'.

The exponent of approximation (2) is also a lower bound:

$$r_{(2)} = (s + s') \sin(\gamma/2), \tag{A.19}$$

$$r_{(2)}^2 = (s^2 + s'^2 + 2ss') \frac{1 - \cos \gamma}{2}, \tag{A.20}$$

$$r_{(2)}^2 = s^2 + s'^2 - 2ss' \cos \gamma - \frac{1}{2}(s - s')^2 (1 + \cos \gamma) \leq s^2 + s'^2 - 2ss' \cos \gamma = r^2. \tag{A.21}$$

So clearly all three approximations represent an upper bound for the exact integral.

A.4.2. Experimental Data

Anemometers

Figure A.6.: Thies Ultrasonic Anemometer 3D [141].

Figure A.7.: Gill WindObserver II [142].

Leg	Longit. Mean (Gill)	Longit. Std.Dev. (Gill)	Longit. Mean (Thies)	Longit. Std.Dev. (Thies)	Transv. Mean (Gill)	Transv. Std.Dev. (Gill)	Transv. Mean (Thies)	Transv. Std.Dev. (Thies)
7	0.72	0.43	0.78	0.45	-0.15	0.45	0.06	0.42
8	0.80	0.56	0.87	0.64	-0.16	0.67	0.12	0.60
9	1.13	0.66	1.11	0.75	-0.23	0.73	0.31	0.59
10	1.10	0.65	1.17	0.77	-0.31	0.72	0.09	0.55
11	0.61	0.38	0.63	0.40	-0.17	0.42	0.05	0.37
12	0.93	0.52	1.31	0.66	-0.87	0.73	-0.55	0.67
13	0.72	0.49	0.85	0.50	-0.47	0.67	-0.28	0.68
14	0.68	0.54	0.74	0.59	-0.23	0.53	-0.04	0.50
21	0.71	0.74	0.63	0.71	-0.53	0.65	-0.68	0.78
22	1.00	0.83	0.90	0.74	-0.55	0.66	-0.73	0.73
23	0.87	0.75	0.60	0.61	-0.58	0.62	-0.76	0.72
24	0.85	0.70	0.95	0.83	-0.65	0.66	-0.01	0.52
25	0.01	0.53	0.40	0.65	-0.98	0.89	-0.30	0.56
26	0.44	0.56	0.92	0.82	-1.01	0.92	-0.03	0.56
27	0.26	0.63	0.72	0.89	-1.09	0.86	-0.05	0.52
28	0.35	0.58	0.68	0.75	-1.09	0.94	-0.16	0.53
29	-0.48	0.75	-0.54	0.77	0.77	0.87	0.84	0.89

Table A.2.: Mean and standard deviation of longitudinal and transversal wind velocity for Gill (Lsp1) and Thies (Lsp2) units in m/s.

Leg	Distance /m	Longitudinal Cross-Corr.	Transversal Cross-Corr.
7	1.7	0.69	0.70
8	1.0	0.74	0.78
9	3.1	0.62	0.48
10	2.4	0.72	0.66
11	2.0	0.72	0.67
12	2.0	0.64	0.73
13	0.2	0.83	0.90
14	0.2	0.83	0.75
21	1.4	0.75	0.73
22	1.4	0.74	0.63
23	2.7	0.58	0.57
24	3.4	0.59	0.50
25	5.0	0.50	0.43
26	5.0	0.43	0.38
27	5.0	0.50	0.30
28	5.0	0.35	0.18
29	0.2	0.65	0.67

Table A.3.: Longitudinal and transversal cross-correlation coefficient as a function of the distance between measurement locations.

Leg	Temperature Std. Dev. / K (Thies)	Temperature Std. Dev. / K (Gill)
7	0.16	0.23
8	0.13	0.17
9	0.10	0.16
10	0.10	0.15
11	0.09	0.14
12	0.07	0.10
13	0.09	0.12
14	0.09	0.13
21	0.09	0.17
22	0.10	0.14
23	0.11	0.17
24	0.17	0.19
25	0.13	0.20
26	0.08	0.11
27	0.08	0.11
28	0.16	0.17

Table A.4.: Standard deviation of temperature for Thies and Gill unit.

Leg	Distance / m	Cross-Corr.
7	1.7	0.21
8	1.0	0.31
9	3.1	0.22
10	2.4	0.15
11	2.0	0.19
12	2.0	0.29
13	0.2	0.66
14	0.2	0.47
21	1.4	0.43
22	1.4	0.53
23	2.7	0.32
24	3.4	0.24
25	5.0	0.15
26	5.0	0.19
27	5.0	0.18
28	5.0	0.11

Table A.5.: Temperature cross-correlation coefficient as a function of the distance between measurement locations.

Leg	Lsp1 Distance / m	Lsp1 Std. Dev. / μs	Lsp2 Distance / m	Lsp2 Std. Dev. / μs	Lsp2-Lsp1 Std. Dev. / μs
7	9.20	33			
8	9.20	38			
9	9.20	41			
10	9.20	44			
21	8.00	33	9.80	36	20
22	8.00	42	9.80	44	23
23	8.00	40	8.90	35	27
24	8.75	40	14.35	52	27

Table A.6.: Standard deviation of propagation time (columns 3 and 5) and of propagation time difference (column 6) for various measurement setups.

Leg	Measurement Duration / min	Distance Thies-Gill / m	Distance Lsp1-Mic / m	Distance Lsp2-Mic / m	Base Angle Lsp1-Lsp2 / °
1*	1	0.2			
2*	1	0.2			
3*	10	0.2			
4*	1	1.7	9.2		
5*	1	1.7	9.2		
6*	1	1.7	9.2		
7	10	1.7	9.2		
8	10	1.0	9.2		
9	10	3.1	9.2		
10	10	2.4	9.2		
11	10	2.0			
12	10	2.0			
13	10	0.2			
14	30	0.2			
15*	1	1.4	8.0	9.8	13
16*	1	1.4	8.0	9.8	13
17*	1	1.4	8.0	9.8	13
18*	1	1.4	8.0	9.8	13
19*	1	1.4	8.0	9.8	13
20*	1	1.4	8.0	9.8	13
21	10	1.4	8.0	9.8	13
22	10	1.4	8.0	9.8	13
23	10	2.7	8.0	8.9	25
24	10	3.4	8.75	14.35	0
25	10	5.0			
26	10	5.0			
27	10	5.0			
28	30	5.0			
29	30	0.2			

Table A.7.: List of measurement series, * indicates test measurements that have not been used in the final analysis.

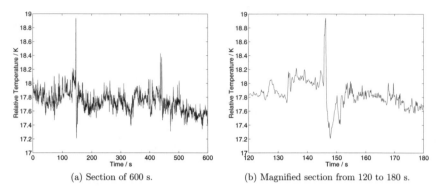

(a) Section of 600 s.   (b) Magnified section from 120 to 180 s.

Figure A.8.: Temperature measurement, raw data, leg 10.

A.4.3. Uncertainty Estimates

Standard Deviation of Propagation Time

An uncertainty estimate can be derived based on standard propagation methods. This involves calculating the first derivative of σ_t with respect to the measurement parameters s, τ, σ_T and α:

$$\delta\sigma_t = \left|\frac{\partial\sigma_t}{\partial s}\right|\delta s + \left|\frac{\partial\sigma_t}{\partial\tau}\right|\delta\tau + \left|\frac{\partial\sigma_t}{\partial\sigma_T}\right|\delta\sigma_T + \left|\frac{\partial\sigma_t}{\partial\alpha}\right|\delta\alpha. \tag{A.22}$$

Given the definition of the standard deviation of the travel time,

$$\sigma_t(s, \tau, \sigma_T, \alpha) = \sqrt{2}|\alpha|\sigma_T\tau\sqrt{s/\tau + e^{-s/\tau} - 1}, \tag{A.23}$$

one can calculate the derivatives as follows:

$$\frac{\partial\sigma_t}{\partial\alpha} = \sqrt{2}\sigma_T\tau\sqrt{s/\tau + e^{-s/\tau} - 1}, \tag{A.24}$$

$$\frac{\partial\sigma_t}{\partial\sigma_T} = \sqrt{2}\alpha\tau\sqrt{s/\tau + e^{-s/\tau} - 1}, \tag{A.25}$$

$$\frac{\partial\sigma_t}{\partial\tau} = \sqrt{2}\alpha\sigma_T\left(\sqrt{s/\tau + e^{-s/\tau} - 1} + \frac{s}{2\tau}\frac{1}{\sqrt{s/\tau + e^{-s/\tau} - 1}}(e^{-s/\tau} - 1)\right), \tag{A.26}$$

$$\frac{\partial\sigma_t}{\partial s} = \frac{1}{\sqrt{2}}\alpha\sigma_T\frac{1}{\sqrt{s/\tau + e^{-s/\tau} - 1}}(1 - e^{-s/\tau}). \tag{A.27}$$

A. Appendix

Standard Deviation of Propagation Time Difference

In the same manner one can calculate an uncertainty estimate for $\sigma_{\Delta t}$ which requires calculating the first derivative with respect to the measurement parameters s_1, s_2, τ, σ_T, α and γ:

$$\delta\sigma_{\Delta t} = \left|\frac{\partial\sigma_{\Delta t}}{\partial s_1}\right|\delta s_1 + \left|\frac{\partial\sigma_{\Delta t}}{\partial s_2}\right|\delta s_2 + \left|\frac{\partial\sigma_{\Delta t}}{\partial\tau}\right|\delta\tau + \left|\frac{\partial\sigma_{\Delta t}}{\partial\gamma}\right|\delta\gamma + \left|\frac{\partial\sigma_{\Delta t}}{\partial\sigma_T}\right|\delta\sigma_T + \left|\frac{\partial\sigma_{\Delta t}}{\partial\alpha}\right|\delta\alpha. \quad \text{(A.28)}$$

The standard deviation is given by

$$\sigma_{\Delta t} = \sqrt{\sigma_{t_1}^2(s_1,\tau,\sigma_T,\alpha) + \sigma_{t_2}^2(s_2,\tau,\sigma_T,\alpha) - 2\alpha^2\sigma_T^2 I(s_1,s_2,\tau,\gamma)}, \quad \text{(A.29)}$$

where σ_{t_1} is the standard deviation of the propagation time along the first path, σ_{t_2} along the second path and I represents the correlation integral discussed earlier.

Calculating the direct derivatives yields

$$\frac{\partial\sigma_{\Delta t}}{\partial s_1} = \frac{1}{2\sigma_{\Delta t}}\left(2\sigma_{t_1}\frac{\partial\sigma_{t_1}}{\partial s_1} - 2\alpha^2\sigma_T^2\frac{\partial I}{\partial s_1}\right), \quad \text{(A.30)}$$

$$\frac{\partial\sigma_{\Delta t}}{\partial s_2} = \frac{1}{2\sigma_{\Delta t}}\left(2\sigma_{t_2}\frac{\partial\sigma_{t_2}}{\partial s_2} - 2\alpha^2\sigma_T^2\frac{\partial I}{\partial s_2}\right), \quad \text{(A.31)}$$

$$\frac{\partial\sigma_{\Delta t}}{\partial\tau} = \frac{1}{2\sigma_{\Delta t}}\left(2\sigma_{t_1}\frac{\partial\sigma_{t_1}}{\partial\tau} + 2\sigma_{t_2}\frac{\partial\sigma_{t_2}}{\partial\tau} - 2\alpha^2\sigma_T^2\frac{\partial I}{\partial\tau}\right), \quad \text{(A.32)}$$

$$\frac{\partial\sigma_{\Delta t}}{\partial\gamma} = -\frac{\alpha^2\sigma_T^2}{\sigma_{\Delta t}}\frac{\partial I}{\partial\gamma}, \quad \text{(A.33)}$$

$$\frac{\partial\sigma_{\Delta t}}{\partial\alpha} = \frac{1}{2\sigma_{\Delta t}}\left(2\sigma_{t_1}\frac{\partial\sigma_{t_1}}{\partial\alpha} + 2\sigma_{t_2}\frac{\partial\sigma_{t_2}}{\partial\alpha} - 4\alpha\sigma_T^2 I\right), \quad \text{(A.34)}$$

$$\frac{\partial\sigma_{\Delta t}}{\partial\sigma_T} = \frac{1}{2\sigma_{\Delta t}}\left(2\sigma_{t_1}\frac{\partial\sigma_{t_1}}{\partial\sigma_T} + 2\sigma_{t_2}\frac{\partial\sigma_{t_2}}{\partial\sigma_T} - 4\alpha^2\sigma_T I\right). \quad \text{(A.35)}$$

Since analytical expression are only available for the linear approximation of the correlation integral I, these will be used instead. This means in the case $s_2 \geq s_1/v$ one has $I = L_1 + L_2 + L_3$, whereas for $s_1/v \geq s_2$ one has $I = K_1 + K_2 + K_3 + K_4$. Now the derivatives of these summands with respect to s_1, s_2, τ and γ can be computed:

$$\frac{\partial L_1}{\partial\tau} = \frac{2\tau}{\cos\gamma}\left[\frac{1}{v\cos\gamma - 1}\left(e^{s_1(v\cos\gamma-1)/\tau} - 1\right) + e^{-s_1/\tau} - 1\right]$$
$$- \frac{s_1}{\cos\gamma}\left(e^{s_1(v\cos\gamma-1)/\tau} - e^{-s_1/\tau}\right), \quad \text{(A.36)}$$

$$\frac{\partial L_1}{\partial s_1} = \frac{\tau}{\cos\gamma}\left(e^{s_1(v\cos\gamma-1)/\tau} - e^{-s_1/\tau}\right), \quad \text{(A.37)}$$

$$\frac{\partial L_1}{\partial s_2} = 0,\qquad\qquad(A.38)$$

$$\frac{\partial L_1}{\partial \gamma} = \frac{\tau^2 \sin\gamma}{\cos^2\gamma}\left[\frac{1}{v\cos\gamma - 1}\left(e^{s_1(v\cos\gamma-1)/\tau} - 1\right) + e^{-s_1/\tau} - 1\right] +$$

$$\frac{\tau^2}{\cos^2\gamma}\left(\frac{\partial v}{\partial \gamma}\cos\gamma - v\sin\gamma\right)\left[\frac{1}{v\cos\gamma - 1}\frac{s_1}{\tau}e^{s_1(v\cos\gamma-1)/\tau}\right.$$

$$\left.- \frac{1}{(v\cos\gamma - 1)^2}\left(e^{s_1(v\cos\gamma-1)/\tau} - 1\right)\right].\quad(A.39)$$

Here, as well as in the following it was used

$$\frac{\partial v}{\partial \gamma} = \frac{-\cos\frac{\gamma}{2}}{2(\cos\gamma + \sin\frac{\gamma}{2})} - \frac{(1 - \sin\frac{\gamma}{2})(-\sin\gamma + \cos\frac{\gamma}{2})}{2(\cos\gamma + \sin\frac{\gamma}{2})^2}.\qquad(A.40)$$

The derivatives for L_2 are:

$$\frac{\partial L_2}{\partial \tau} = \frac{2\tau}{\sin^2\frac{\gamma}{2}}\left[\frac{1}{1 + 1/v}\left(e^{-s_1\sin\frac{\gamma}{2}(1+1/v)/\tau} - 1\right) - \frac{1}{1 + v}e^{-s_1\sin\frac{\gamma}{2}(1+v)/\tau} - 1\right)\right]$$

$$+ \frac{s_1}{\sin\frac{\gamma}{2}}\left[e^{-s_1\sin\frac{\gamma}{2}(1+1/v)/\tau} - e^{-s_1\sin\frac{\gamma}{2}(1+v)/\tau}\right],\quad(A.41)$$

$$\frac{\partial L_2}{\partial s_1} = \frac{-\tau}{\sin\frac{\gamma}{2}}\left(e^{-s_1\sin\frac{\gamma}{2}(1+1/v)/\tau} - e^{-s_1\sin\frac{\gamma}{2}(1+v)/\tau}\right),\qquad(A.42)$$

$$\frac{\partial L_2}{\partial s_2} = 0,\qquad\qquad(A.43)$$

$$\frac{\partial L_2}{\partial \gamma} = \frac{-\tau^2\cos\frac{\gamma}{2}}{\sin^3\frac{\gamma}{2}}\left[\frac{1}{1 + 1/v}\left(e^{-s_1\sin\frac{\gamma}{2}(1+1/v)/\tau} - 1\right) - \frac{1}{1 + v}\left(e^{-s_1\sin\frac{\gamma}{2}(1+v)/\tau} - 1\right)\right]$$

$$+ \frac{\tau^2}{\sin^2\frac{\gamma}{2}}\left\{\frac{1}{(v + 1)^2}\frac{\partial v}{\partial \gamma}\left[e^{-s_1\sin\frac{\gamma}{2}(1+1/v)/\tau} - 1\right]\right.$$

$$- \left[\frac{s_1}{\tau}e^{-s_1\sin\frac{\gamma}{2}(1+1/v)/\tau}\left(\frac{1}{2}\cos\frac{\gamma}{2} - \frac{1}{v(v + 1)}\frac{\partial v}{\partial \gamma}\sin\frac{\gamma}{2}\right)\right]$$

$$+ \frac{1}{(1 + v)^2}\frac{\partial v}{\partial \gamma}\left[e^{-s_1\sin\frac{\gamma}{2}(1+v)/\tau} - 1\right]$$

$$\left.+ \left[\frac{s_1}{\tau}e^{-s_1\sin\frac{\gamma}{2}(1+v)/\tau}\left(\frac{1}{2}\cos\frac{\gamma}{2} + \frac{1}{1 + v}\frac{\partial v}{\partial \gamma}\sin\frac{\gamma}{2}\right)\right]\right\}.\quad(A.44)$$

A. Appendix

The derivatives for L_3 are:

$$\frac{\partial L_3}{\partial \tau} = -2\tau \left[\frac{e^{-s_2/\tau}}{\cos \gamma} \left(e^{s_1 \cos \gamma/\tau} - 1 \right) - \frac{1}{\cos \gamma - 1/v} \left(e^{s_1 (\cos \gamma - 1/v)/\tau} - 1 \right) \right]$$
$$+ e^{-s_2/\tau} \left[-\frac{s_2}{\cos \gamma} \left(e^{s_1 \cos \gamma/\tau} - 1 \right) + s_1 e^{s_1 \cos \gamma/\tau} \right] - s_1 e^{s_1 (\cos \gamma - 1/v)/\tau}, \quad (A.45)$$

$$\frac{\partial L_3}{\partial s_1} = -\tau \left(e^{(s_1 \cos \gamma - s_2)/\tau} - e^{s_1 (\cos \gamma - 1/v)/\tau} \right), \quad (A.46)$$

$$\frac{\partial L_3}{\partial s_2} = \frac{\tau}{\cos \gamma} e^{-s_2/\tau} \left(e^{s_1 \cos \gamma/\tau} - 1 \right), \quad (A.47)$$

$$\frac{\partial L_3}{\partial \gamma} = -\tau^2 \left\{ \frac{\sin \gamma}{\cos \gamma} e^{-s_2/\tau} \left[\frac{1}{\cos \gamma} \left(e^{s_1 \cos \gamma/\tau} - 1 \right) - \frac{s_1}{\tau} e^{(s_1 \cos \gamma)/\tau} \right] + \frac{1}{\cos \gamma - 1/v} \right.$$
$$\left. \times \left(\frac{1}{v^2} \frac{\partial v}{\partial \gamma} - \sin \gamma \right) \left[\frac{1}{(\cos \gamma - 1/v)} \left(e^{s_1 (\cos \gamma - 1/v)/\tau} - 1 \right) - \frac{s_1}{\tau} e^{s_1 (\cos \gamma - 1/v)/\tau} \right] \right\}. \quad (A.48)$$

The derivatives for K_1 are:

$$\frac{\partial K_1}{\partial \tau} = \frac{\partial L_1}{\partial \tau}, \quad (A.49)$$

$$\frac{\partial K_1}{\partial s_1} = \frac{\partial L_1}{\partial s_1}, \quad (A.50)$$

$$\frac{\partial K_1}{\partial s_2} = \frac{\partial L_1}{\partial s_2}, \quad (A.51)$$

$$\frac{\partial K_1}{\partial \gamma} = \frac{\partial L_1}{\partial \gamma}. \quad (A.52)$$

The derivatives for K_2 are:

$$\frac{\partial K_2}{\partial \tau} = \frac{2\tau}{\sin^2 \frac{\gamma}{2}} \left[\frac{1}{1+1/v} \left(e^{-s_2(v+1) \sin \frac{\gamma}{2}/\tau} - 1 \right) - \frac{1}{1+v} \left(e^{-s_2 v(1+v) \sin \frac{\gamma}{2}/\tau} - 1 \right) \right]$$
$$+ \frac{s_2 v}{\sin \frac{\gamma}{2}} \left[e^{-s_2(v+1) \sin \frac{\gamma}{2}/\tau} - e^{-s_2 v(1+v) \sin \frac{\gamma}{2}/\tau} \right], \quad (A.53)$$

$$\frac{\partial K_2}{\partial s_1} = 0, \quad (A.54)$$

$$\frac{\partial K_2}{\partial s_2} = \frac{-\tau v}{\sin \frac{\gamma}{2}} \left[e^{-s_2(v+1) \sin \frac{\gamma}{2}/\tau} - e^{-s_2 v(1+v) \sin \frac{\gamma}{2}/\tau} \right], \quad (A.55)$$

$$\frac{\partial K_2}{\partial \gamma} = \frac{-\tau^2 \cos \frac{\gamma}{2}}{\sin^3 \frac{\gamma}{2}} \left[\frac{1}{1+1/v} \left(e^{-s_2 \sin \frac{\gamma}{2}(v+1)/\tau} - 1 \right) - \frac{1}{1+v} \left(e^{-s_2 \sin \frac{\gamma}{2}v(1+v)/\tau} - 1 \right) \right]$$

$$+ \frac{\tau^2}{\sin^2 \frac{\gamma}{2}} \left\{ \frac{1}{(v+1)^2} \frac{\partial v}{\partial \gamma} \left[e^{-s_2 \sin \frac{\gamma}{2}(v+1)/\tau} - 1 \right] \right.$$

$$- \left[\frac{s_2 v}{\tau} e^{-s_2 \sin \frac{\gamma}{2}(v+1)/\tau} \left(\frac{1}{2} \cos \frac{\gamma}{2} + \frac{1}{v+1} \frac{\partial v}{\partial \gamma} \sin \frac{\gamma}{2} \right) \right]$$

$$+ \frac{1}{(1+v)^2} \frac{\partial v}{\partial \gamma} \left[e^{-s_2 \sin \frac{\gamma}{2}v(1+v)/\tau} - 1 \right]$$

$$\left. + \left[\frac{s_2 v}{\tau} e^{-s_2 \sin \frac{\gamma}{2}v(1+v)/\tau} \left(\frac{1}{2} \cos \frac{\gamma}{2} + \frac{2v+1}{v(1+v)} \frac{\partial v}{\partial \gamma} \sin \frac{\gamma}{2} \right) \right] \right\}. \quad (A.56)$$

The derivatives for K_3 are:

$$\frac{\partial K_3}{\partial \tau} = -2\tau \left[\frac{e^{-s_2/\tau}}{\cos \gamma} \left(e^{s_2 v \cos \gamma/\tau} - 1 \right) - \frac{1}{\cos \gamma - 1/v} \left(e^{s_2 v (\cos \gamma - 1/v)/\tau} - 1 \right) \right]$$

$$+ e^{-s_2/\tau} \left[-\frac{s_2}{\cos \gamma} \left(e^{s_2 v \cos \gamma/\tau} - 1 \right) + s_2 v e^{s_2 v \cos \gamma/\tau} \right] - s_2 v e^{s_2 v (\cos \gamma - 1/v)/\tau}, \quad (A.57)$$

$$\frac{\partial K_3}{\partial s_1} = 0, \quad (A.58)$$

$$\frac{\partial K_3}{\partial s_2} = -\tau v \left(e^{(s_2 v \cos \gamma - s_2)/\tau} - e^{s_2 v (\cos \gamma - 1/v)/\tau} \right) + \frac{\tau}{\cos \gamma} e^{-s_2/\tau} \left(e^{s_2 v \cos \gamma/\tau} - 1 \right), \quad (A.59)$$

$$\frac{\partial K_3}{\partial \gamma} = -\tau^2 \left[e^{-s_2/\tau} \frac{\sin \gamma}{\cos^2 \gamma} \left(e^{s_2 v \cos \gamma/\tau} - 1 \right) + \frac{e^{-s_2/\tau}}{\cos \gamma} \frac{s_2}{\tau} e^{s_2 v \cos \gamma/\tau} \left(\frac{\partial v}{\partial \gamma} \cos \gamma - v \sin \gamma \right) \right.$$

$$+ \frac{1}{(\cos \gamma - 1/v)^2} \left(\frac{1}{v^2} \frac{\partial v}{\partial \gamma} - \sin \gamma \right) \left(e^{s_2 v (\cos \gamma - 1/v)/\tau} - 1 \right)$$

$$\left. - \frac{1}{\cos \gamma - 1/v} \frac{s_2}{\tau} e^{s_2 v (\cos \gamma - 1/v)/\tau} \left(\frac{\partial v}{\partial \gamma} \cos \gamma - v \sin \gamma \right) \right]. \quad (A.60)$$

A. Appendix

The derivatives for K_4 are:

$$\frac{\partial K_4}{\partial \tau} = \frac{2\tau}{\sin^2 \frac{\gamma}{2}} \left[e^{-s_2 \sin \frac{\gamma}{2}/\tau} \left(e^{-s_1 \sin \frac{\gamma}{2}/\tau} - e^{-s_2 v \sin \frac{\gamma}{2}/\tau} \right) \right.$$

$$\left. - \frac{1}{1+v} \left(e^{-s_1(1+v)\sin \frac{\gamma}{2}/\tau} - e^{-s_2 v(v+1)\sin \frac{\gamma}{2}/\tau} \right) \right]$$

$$+ \frac{1}{\sin \frac{\gamma}{2}} \left[e^{-s_2 \sin \frac{\gamma}{2}/\tau} \left(s_2 e^{-s_1 \sin \frac{\gamma}{2}/\tau} - s_2 e^{-s_2 v \sin \frac{\gamma}{2}/\tau} + s_1 e^{-s_1 \sin \frac{\gamma}{2}/\tau} - s_2 v e^{-s_2 v \sin \frac{\gamma}{2}/\tau} \right) \right.$$

$$\left. - s_1 e^{-s_1(1+v)\sin \frac{\gamma}{2}/\tau} + s_2 v e^{-s_2 v(v+1)\sin \frac{\gamma}{2}/\tau} \right], \quad \text{(A.61)}$$

$$\frac{\partial K_4}{\partial s_1} = \frac{-\tau}{\sin \frac{\gamma}{2}} \left(e^{-(s_1+s_2)\sin \frac{\gamma}{2}/\tau} - e^{-s_1(1+v)\sin \frac{\gamma}{2}/\tau} \right), \quad \text{(A.62)}$$

$$\frac{\partial K_4}{\partial s_2} = \frac{-\tau}{\sin \frac{\gamma}{2}} \left[e^{-(s_1+s_2)\sin \frac{\gamma}{2}/\tau} - (1+v)e^{-s_2(1+v)\sin \frac{\gamma}{2}/\tau} + v e^{-s_2 v(v+1)\sin \frac{\gamma}{2}/\tau} \right], \quad \text{(A.63)}$$

$$\frac{\partial K_4}{\partial \gamma} = \frac{-\tau^2 \cos \frac{\gamma}{2}}{\sin^3 \frac{\gamma}{2}} \left[e^{-s_2 \sin \frac{\gamma}{2}/\tau} \left(e^{-s_1 \sin \frac{\gamma}{2}/\tau} - e^{-s_2 v \sin \frac{\gamma}{2}/\tau} \right) \right.$$

$$\left. - \frac{1}{1+v} \left(e^{-s_1(1+v)\sin \frac{\gamma}{2}/\tau} - e^{-s_2 v(v+1)\sin \frac{\gamma}{2}/\tau} \right) \right]$$

$$+ \frac{\tau}{\sin^2 \frac{\gamma}{2}} \left\{ -\frac{1}{2} s_2 \cos \frac{\gamma}{2} e^{-s_2 \sin \frac{\gamma}{2}/\tau} \left(e^{-s_1 \sin \frac{\gamma}{2}/\tau} - e^{-s_2 v \sin \frac{\gamma}{2}/\tau} \right) \right.$$

$$+ e^{-s_2 \sin \frac{\gamma}{2}/\tau} \left[-\frac{1}{2} s_1 \cos \frac{\gamma}{2} e^{-s_1 \sin \frac{\gamma}{2}/\tau} + s_2 \left(\frac{\partial v}{\partial \gamma} \sin \frac{\gamma}{2} + \frac{1}{2} v \cos \frac{\gamma}{2} \right) e^{-s_2 v \sin \frac{\gamma}{2}/\tau} \right]$$

$$+ \frac{1}{(1+v)^2} \frac{\partial v}{\partial \gamma} \left(e^{-s_1(1+v)\sin \frac{\gamma}{2}/\tau} - e^{-s_2 v(v+1)\sin \frac{\gamma}{2}/\tau} \right)$$

$$- \frac{1}{1+v} \left[-s_1 e^{-s_1(1+v)\sin \frac{\gamma}{2}/\tau} \left(\frac{\partial v}{\partial \gamma} \sin \frac{\gamma}{2} + \frac{1}{2}(1+v)\cos \frac{\gamma}{2} \right) \right.$$

$$\left. \left. + s_2 e^{-s_2 v(1+v)\sin \frac{\gamma}{2}/\tau} \left(\frac{\partial v}{\partial \gamma}(2v+1)\sin \frac{\gamma}{2} + \frac{1}{2}v(1+v)\cos \frac{\gamma}{2} \right) \right] \right\}. \quad \text{(A.64)}$$

A. Appendix

A.4.4. Comparison with Existing Work

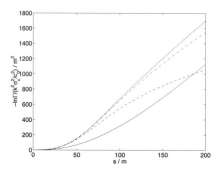

Figure A.9.: Coherence factor using Gaussian spectrum model, 315 Hz (- -), 1000 Hz (-.-), 3150 Hz ($\cdots$), and exponential decay model ($-$), $\gamma = 5°$, $\tau = 5$ m, wind velocity only.

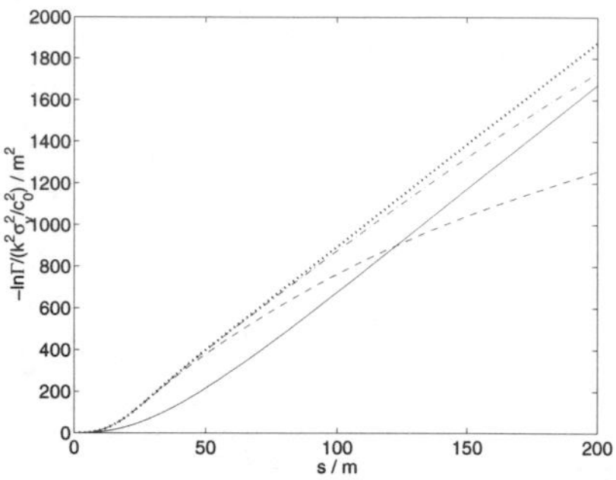

Figure A.10.: Coherence factor using Gaussian spectrum model, 315 Hz (- -), 1000 Hz (-.-), 3150 Hz ($\cdots$), and exponential decay model ($-$), $\gamma = 15°$, $\tau = 5$ m, wind velocity only.

Figures A.9 and A.10 show the same data as Figure 4.25 but for a different geometrical setup, namely with respect to angle γ. The same relationship but for temperature data is shown in Figure A.11.

Complementing Figure 4.26, Figure A.12 shows the equivalent temperature quantities $\langle \phi^2 \rangle \sim 1 + M(D)$ and $\langle \chi^2 \rangle \sim 1 - M(D)$.

Finally, Figures A.13 and A.14 show two exact Gaussian spectrum curves and the $D \gg 1$ asymptotes of the Gaussian and von Karman model. The exact curves assume the same slope as the asymptotes when $s \gg \lambda$, for both wind velocity and temperature. However, for smaller s the slope is about twice the asymptotic slope.

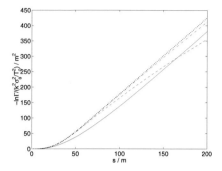

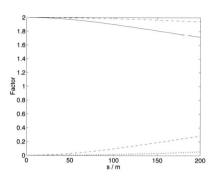

Figure A.11.: Coherence factor using Gaussian spectrum model, 315 Hz (- -), 1000 Hz (-.-), 3150 Hz (···), and exponential decay model (−), $\gamma = 10°$, $\tau = 5$ m, temperature only.

Figure A.12.: Relative strength of fluctuations according to Gaussian spectrum model, phase 315 Hz (−) and 1000 Hz (-.-), amplitude 315 Hz (- -) and 1000 Hz (···), $\tau = 5$ m, temperature only.

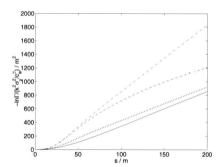

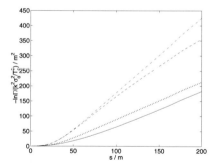

Figure A.13.: Coherence factor using Gaussian spectrum model, 315 Hz (- -), 3150 Hz (-.-), and asymptotes $D \gg 1$ Gaussian spectrum model (···) and von Karman model (−), $\gamma = 10°$, $\tau = 5$ m, wind velocity only.

Figure A.14.: Coherence factor using Gaussian spectrum model, 315 Hz (- -), 3150 Hz (-.-), and asymptotes $D \gg 1$ Gaussian spectrum model (···) and von Karman model (−), $\gamma = 10°$, $\tau = 5$ m, temperature only.

A. Appendix

Bibliography

[1] L. Kinsler, A. Frey, A. Coppens, J. Sanders, *Fundamentals of Acoustics*, 4th ed. (Wiley, New York, 2000).

[2] E. Skudrzyk, *Die Grundlagen der Akustik*, 1. Auflage (Springer, Wien, 1954).

[3] P. M. Morse, K. U. Ingard, *Theoretical Acoustics*, 1st ed., with errata page (Princeton University Press, Princeton, 1986).

[4] R. B. Lindsay (Ed.), *Physical Acoustics*, 1st. ed. (Dowden, Hutchinson & Ross, Stroudsburg, 1974).

[5] M. Vorländer, *Auralization*, 1st ed. (Springer, Berlin, 2008).

[6] G. Ballou (Ed.), *Handbook for Sound Engineers*, 4th ed. (Focal Press, Burlington, 2008).

[7] M. R. Schroeder, "Frequency-Correlation Functions of Frequency Responses in Rooms", *J. Acoust. Soc. Am.*, vol. 34, no. 12, pp. 1819-1823 (1962): M. R. Schroeder, "The 'Schroeder frequency' revisited", *J. Acoust. Soc. Am.*, vol. 99, no. 5, pp. 3240-3241 (1996).

[8] R. Green, T. Holman, "First Results from a Large-Scale Measurement Program for Home Theaters", presented at the 129th Convention of the Audio Engineering Society, *J. Audio Eng. Soc. (Abstracts)*, convention paper 8310 (2010 Nov.).

[9] L. Cremer, H. A. Müller, *Die wissenschaftlichen Grundlagen der Raumakustik*, Band I, 2. Auflage (S. Hirzel, Stuttgart, 1978): L. Cremer, H. A. Müller, *Die wissenschaftlichen Grundlagen der Raumakustik*, Band II, 2. Auflage (S. Hirzel, Stuttgart, 1976).

[10] H. Kuttruff, "A simple iteration scheme for the computation of decay constants in enclosures with diffusely reflecting boundaries", *J. Acoust. Soc. Am.*, vol. 98, pp. 288-293 (1995).

[11] M. Vorländer, "Simulation of the transient and steady state sound propagation in rooms using a new combined sound particle - image source algorithm", *J. Acoust. Soc. Am.*, vol. 86, pp. 172 (1989).

[12] R. Heinz, "Binaural Room Simulation Based on the Image Source Model with Addition of Statistical Methods to Include the Diffuse Sound Scattering of Walls

and to Predict the Reverberant Tail", *Appl. Ac.*, vol. 38, pp. 145 (1993): R. Heinz, *Entwicklung und Beurteilung von computergestützten Methoden zur binauralen Raumsimulation*, PhD thesis, Institute of Technical Acoustics, RWTH Aachen University (1994).

[13] U. M. Stephenson, *Beugungssimulation ohne Rechenzeitexplosion: Die Methode der quantisierten Pyramidenstrahlen - ein neues Berechnungsverfahren für Raumakustik und Lärmimmissionsprognose*, PhD thesis, Institute of Technical Acoustics, RWTH Aachen University (2004).

[14] I. Bork, "A Comparison of Room Simulation Software - the 2nd Round Robin on Room Acoustical Computer Simulation", *Acustica united with Acta Acustica*, vol. 84, pp. 943 (2000).

[15] O. Schmitz, S. Feistel, W. Ahnert, M. Vorländer, "Grundlagen raumakustischer Rechenverfahren und ihre Validierung", *Fortschritte der Akustik - DAGA '01*, pp. 24-25 (2001).

[16] S. Feistel, W. Ahnert, A. Miron, O. Schmitz, "Improved Methods for Calculating Room Impulse Responses with EASE 4.2 AURA", presented at the 19th International Congress on Acoustics (Barcelona, Spain, 2007).

[17] EASE software, http://ease.afmg.eu: CATT-Acoustic software, http://www.catt.se: ODEON software, http://www.odeon.dk.

[18] O. Schmitz, *Entwicklung und Programmierung eines Hybrid-Verfahrens zur Simulation der Schallübertragung in Räumen*, Diploma thesis, Institute of Technical Acoustics, RWTH Aachen University (1997).

[19] U. P. Svensson, "The inclusion of diffraction effects in room acoustical modeling", *J. Acoust. Soc. Am.*, vol. 129, pp. 2365-2365 (2011): R. Torres, U. P. Svensson, M. Kleiner, "Computation of edge diffraction for more accurate room acoustics auralization", *J. Acoust. Soc. Am.*, vol. 109, pp. 600-610 (2001): U. P. Svensson, R. I. Fred, J. Vanderkooy, "An analytic secondary source model of edge diffraction impulse responses", *J. Acoust. Soc. Am.*, vol. 106, pp. 2331-2344 (1999).

[20] L.C. Wrobel, *The Boundary Element Method*, vol. 1, 1st ed. (John Wiley & Sons, Chichester, 2007).

[21] S. M. Kirkup, *The Boundary Element Method in Acoustics*, 2nd ed., www.boundary-element-method.com.

[22] G. Bartsch, *Effiziente Methoden für die niederfrequente Schallfeldsimulation*, 1st ed. (Driesen, Taunusstein, 2003).

[23] H. Schmalle, D. Noy, S. Feistel, G. Hauser, W. Ahnert, J. Storyk, "Accurate Acoustic Modeling of Small Rooms", presented at the 131st Convention of the

Audio Engineering Society, *J. Audio Eng. Soc. (Abstracts)*, convention paper 8457 (2011 Oct.).

[24] J. Häggblad, B. Enquist, "Consistent modeling of boundaries in acoustic finite-difference time-domain simulations", *J. Acoust. Soc. Am.*, vol. 132, pp 1303-1310 (2012).

[25] COMSOL Software, http://www.comsol.com.

[26] ISO 3382-1:2009-06, "Acoustics - Measurement of room acoustic parameters - Part 1: Performance spaces": ISO 3382-2:2008-06, "Acoustics - Measurement of room acoustic parameters - Part 2: Reverberation time in ordinary rooms".

[27] IEC 60268-16:2011, "Sound system equipment - Part 16: Objective rating of speech intelligibility by speech transmission index".

[28] W. Ahnert, F. Steffen, *Sound Reinforcement Engineering*, 1st ed. (E & FN SPON, London, 1999).

[29] L. L. Beranek, *Acoustics*, 1993 edition (Acoustical Society of America, New York, 1996).

[30] Y. Ando, *Architectural Acoustics*, 1st ed. (Springer, New York, 1998).

[31] T. Lentz, *Binaural technology for virtual reality*, PhD thesis, Institute of Technical Acoustics, RWTH Aachen University (1993).

[32] W. Ahnert, R. Feistel, "Binaural auralization from a sound system simulation program", presented at the 91st Convention of the Audio Engineering Society, *J. Audio Eng. Soc. (Abstracts)*, vol. 39, pp. 996, convention paper 3127 (1991 Oct.).

[33] M. Aretz, R. Nöthen, M. Vorländer, D. Schröder: "Combined broadband impulse responses using FEM and hybrid ray-based methods", *Proc. of the EAA Symposium on Auralization* (2009 June).

[34] W. Reichardt, *Grundlagen der Elektroakustik*, 1. Auflage, (Akademische Verlagsgesellschaft Geest & Portig, Leipzig, 1952).

[35] J. H. Rindel, F. Otondo, C. L. Christensen, "Sound Source Representation for Auralization", *International Symposium on Room Acoustics*, Kyoto (April 2004).

[36] D. A. Bies, C. H. Hansen, *Engineering Noise Control - Theory and Practice*, 3rd ed. (Spon Press, London, 2003).

[37] C. M. Harris (Ed.), *Handbook of Acoustical Measurements and Noise Control*, reprint of 3rd ed. (Acoustical Society of America, Melville, 1998).

[38] M. Urban, C. Heil, P. Bauman, "Wavefront Sculpture Technology", *J. Audio Eng. Soc.*, vol. 51, no. 10, pp. 912-932 (2003 Oct.).

[39] M. S. Ureda, "Analysis of Loudspeaker Line Arrays", *J. Audio Eng. Soc.*, vol. 52, no. 5, pp. 467-495 (2004 May).

[40] G. de Vries, G. v. Beuningen, "Concepts and Applications of Directivity Controlled Loudspeaker Arrays", *J. Acoust. Soc. Am.*, vol. 101, pp. 3051-3051 (1997).

[41] J. van der Werff, "Design and Implementation of a Sound Column with Exceptional Properties", presented at the 96th Convention of the Audio Engineering Society, *J. Audio Eng. Soc. (Abstracts)*, convention paper 3835 (1994 May).

[42] D. Gunness, "Improving Loudspeaker Transient Response with Digital Signal Processing", presented at the 119th Convention of the Audio Engineering Society, *J. Audio Eng. Soc. (Abstracts)*, convention paper 6590 (2005 Oct.).

[43] G. Behler, A. Goert, J. Kleber, M. Makarski, S. Müller, R. Thaden, "A Loudspeaker Management System With FIR/IIR Filtering", presented at the AES 32nd International Conference, convention paper 23 (2007 Sep.).

[44] F. Rumsey, "DSP in Loudspeakers", *J. Audio Eng. Soc.*, vol. 56, no. 1/2, pp. 65-72 (2008 Jan.).

[45] A. Thompson, "Improved Methods for Controlling Touring Loudspeaker Arrays", presented at the 127th Convention of the Audio Engineering Society, *J. Audio Eng. Soc. (Abstracts)*, convention paper 7828 (2009 Oct.).

[46] D. G. Meyer, "Computer Simulation of Loudspeaker Directivity", *J. Audio Eng. Soc.*, vol. 32, no. 5, pp. 294-315 (1984 May).

[47] K. D. Jacob, T. K. Birkle, "Prediction of the Full-Space Directivity Characteristics of Loudspeaker Arrays", *J. Audio Eng. Soc.*, vol. 38, no. 4, pp. 250-259 (1990 Apr.).

[48] S. Feistel, W. Ahnert, S. Bock, "New Data Format to Describe Complex Sound Sources", presented at the 119th Convention of the Audio Engineering Society, *J. Audio Eng. Soc. (Abstracts)*, convention paper 6631 (2005 Dec.).

[49] S. Feistel, W. Ahnert, "Modeling of Loudspeaker Systems Using High-Resolution Data", *J. Audio Eng. Soc.*, vol. 55, no. 7/8, pp. 571-597 (2007 July).

[50] S. Feistel, A. Thompson, W. Ahnert, "Methods and Limitations of Line Source Simulation", *J. Audio Eng. Soc.*, vol. 57, no. 6, pp. 379-402 (2009 June).

[51] AES56-2008, "AES standard on acoustics - Sound source modeling - Loudspeaker polar radiation measurements".

[52] M. Makarski, "Simulation of Harmonic Distortion in Horns Using an Extended BEM Postprocessing", presented at the 119th Convention of the Audio Engineering Society, *J. Audio Eng. Soc. (Abstracts)*, convention paper 6591 (2005 Oct.).

[53] S. M. Kirkup, A. Thompson, "Simulation of the Acoustic Field Produced by Cavities Using the Boundary Element-Rayleigh Integral Method and Its Application to a Horn Loudspeaker", presented at the International Congress on Sound and Vibration (St Petersburgh, Russia, 2004).

[54] E. Start, "Simulation and application of beam-shaped subwoofer arrays", presented at the 26th Reproduced Sound, *Proceedings of the Institute of Acoustics*, vol. 32, no. 5, pp. 79 (2010).

[55] J. Panzer, "Coupling Lumped and Boundary Element Methods Using Superposition", presented at the 133th Convention of the Audio Engineering Society, *J. Audio Eng. Soc. (Abstracts)*, convention paper 8723 (2012 Oct.).

[56] A. Pietrzyk, M. Kleiner, "The Application of the Finite-Element Method to the Prediction of Soundfields of Small Rooms at Low Frequencies", presented at the 102nd Convention of the Audio Engineering Society, *J. Audio Eng. Soc. (Abstracts)*, convention paper 4423 (1997 Mar.).

[57] R. Beigelbeck, H. Pichler, "FEM-Analysis of Beam Forming in Safety Relevant Workspaces", presented at the 122nd Convention of the Audio Engineering Society, *J. Audio Eng. Soc. (Abstracts)*, convention paper 7061 (2007 May).

[58] ISO 9613-1:1993-06, "Acoustics; attenuation of sound during propagation outdoors; part 1: calculation of the absorption of sound by the atmosphere": ISO 9613-2:1996-12, "Acoustics - Attenuation of sound during propagation outdoors - Part 2: General method of calculation".

[59] D. G. Albert, *Past research on sound propagation through forests*, Cold Regions Research and Engineering Laboratory (October 2004).

[60] E. Shabalina, M. Vorländer, "The propagation of low frequency sound through an audience", presented at the 5th Congress of Alps-Adria Acoustics Association (Petrane, Croatia, 2012).

[61] E. M. Salomons, *Computational Atmospheric Acoustics*, 1st ed. (Kluwer Academic, Dordrecht, 2001).

[62] K. Attenborough, K. M. Li, K. Horoshenkov, *Predicting Outdoor Sound*, 1st ed. (Taylor & Francis, Oxon, 2007).

[63] D. A. Bohn, "Environmental Effects on the Speed of Sound", *J. Audio Eng. Soc.*, vol. 36, no. 4, pp. 223-231 (1988 Apr.).

[64] C. Hughes, *How Accurate is Your Directivity Data?*, technical white paper, www.excelsior-audio.com (2005 Oct.).

[65] V. E. Ostashev, *Acoustics in Moving Inhomogeneous Media*, 1st ed. (E & FN SPON, London, 1997).

[66] IEC 60268-4:2010, "Sound system equipment - Part 4: Microphones".

[67] AES SC-04-04 Working Group on Microphone Measurement and Characterization of the SC-04 Subcommittee on Acoustics: AES-X085 development project, "Detailed Professional Microphone Specifications".

[68] G. Torio, "Understanding the Transfer Functions of Directional Condenser Microphones in Response to Different Sound Sources", presented at the 105th Convention of the Audio Engineering Society, *J. Audio Eng. Soc. (Abstracts)*, convention paper 4800 (1998 Sep.): G. Torio, J. Segota, "Unique Directional Properties of Dual-Diaphragm Microphones", presented at the 109th Convention of the Audio Engineering Society, *J. Audio Eng. Soc. (Abstracts)*, convention paper 5179 (2000 Sep.).

[69] N. E. Milanov, B. E. Milanova, "Proximity Effect of microphone", presented at the 110th Convention of the Audio Engineering Society, *J. Audio Eng. Soc. (Abstracts)*, convention paper 5342 (2001 May).

[70] H. Lehnert, J. Blauert, "Aspects of Auralization in Binaural Room Simulation", presented at the 93rd Convention of the Audio Engineering Society, *J. Audio Eng. Soc. (Abstracts)*, convention paper 3390 (1992 Oct.).

[71] W. Gardner, K. Martin, "HRTF measurements of a KEMAR", *J. Acoust. Soc. Am. (Technical Notes and Research Briefs)*, vol. 97, no. 6, pp. 3907-3908 (1995): http://sound.media.mit.edu/resources/KEMAR.html

[72] T. Ajdler, L. Sbaiz, M. Vetterlib, "Dynamic measurement of room impulse responses using a moving microphone", *J. Acoust. Soc. Am.*, vol. 122, pp. 1636-1645 (2007 Sep.).

[73] B. Supper, "Processing and improving a head-related impulse response database for auralization", presented at the 129th Convention of the Audio Engineering Society, *J. Audio Eng. Soc. (Abstracts)*, convention paper 8267 (2010 Nov.).

[74] A. Schmitz, *Naturgetreue Wiedergabe kopfbezogener Schallaufnahmen über zwei Lautsprecher mit Hilfe eines Übersprechkompensators*, PhD thesis, Institute of Technical Acoustics, RWTH Aachen University (1993).

[75] T. Lentz, O. Schmitz, "Realisation of an adaptive cross-talk cancellation system for a moving listener", presented at the AES 21st International Conference, convention paper 134 (2002 June).

[76] AES Staff, "Spatial Audio", *J. Audio Eng. Soc.*, vol. 55, no. 6, pp. 537-541 (2007 June).

[77] IOSONO GmbH, Spatial Audio Processor, http://www.iosono.de.

[78] W. Ahnert, S. Feistel, T. Lentz, C. Moldrzyk, S. Weinzierl, "Head-Tracked Aural-ization of Acoustical Simulation", presented at the 117th Convention of the Audio Engineering Society, *J. Audio Eng. Soc. (Abstracts)*, convention paper 6275 (2004 Oct.).

[79] P. Dross, D. Schröder, M. Vorländer, "A Fast Reverberation Estimator for Virtual Environments", presented at the AES 30th International Conference, convention paper 13 (2007 Mar.).

[80] T. Lentz, "Dynamic Crosstalk Cancellation for Binaural Synthesis in Virtual Re-ality Environments", *J. Audio Eng. Soc.*, vol. 54, no. 4, pp. 283-294 (2006 Apr.).

[81] E. Mommertz, *Untersuchung akustischer Wandeigenschaften und Modellierung der Schallrückwürfe in der binauralen Raumsimulation*, PhD thesis, Institute of Technical Acoustics, RWTH Aachen University (1994).

[82] T. J. Cox, P. DAntonio, *Acoustic Absorbers and Diffusers*, 1st ed. (Spon Press, London 2004).

[83] E. El-Saghir, S. Feistel, "Influence of Ray Angle of Incidence and Complex Reflec-tion Factor on Acoustical Simulation Results", presented at the 116th Convention of the Audio Engineering Society, *J. Audio Eng. Soc. (Abstracts)*, convention pa-per 6171 (2004 July/Aug.): E. El-Saghir, S. Feistel, "Influence of Ray Angle of Incidence and Complex Reflection Factor on Acoustical Simulation Results (Part II)", presented at the 120th Convention of the Audio Engineering Society, *J. Audio Eng. Soc. (Abstracts)*, convention paper 6737 (2006 July/ Aug.).

[84] J. H. Rindel, "Modelling the Angle-Dependent Pressure Reflection Factor", *App. Acous.*, vol. 38, pp. 223-234 (1993).

[85] ISO 354:2003-05, "Acoustics - Measurement of sound absorption in a reverberation room".

[86] E. Mommertz, "Determination of scattering coefficients from the reflection di-rectivity of architectural surfaces", *App. Acous.*, vol. 60, pp. 201-203 (2000): M. Vorländer, E Mommertz, "Definition and measurement of random-incidence scat-tering coefficients", *App. Acous.*, vol. 60, pp. 187-199 (2000).

[87] ISO 17497-1:2004-05, "Acoustics - Sound-scattering properties of surfaces - Part 1: Measurement of the random-incidence scattering coefficient in a reverberation room": ISO 17497-2:2012-05, "Acoustics - Sound-scattering properties of surfaces - Part 2: Measurement of the directional diffusion coefficient in a free field".

[88] M. Bansal, S. Feistel, W. Ahnert, S. Bock, "Investigating the Scattering Behavior of Incident Plane Waves using BEM", presented at the 19th International Congress on Acoustics (Barcelona, Spain, 2007).

[89] F. P. Mechel (Ed.), *Formulas of Acoustics*, 1st ed. (Springer-Verlag, Berlin, 2002).

[90] M. Aretz, M. Vorländer, "Efficient Modelling of Absorbing Boundaries in Room Acoustic FE Simulations", *Acta Acustica united with Acustica*, vol. 96, no. 6, pp. 1042-1050 (2010 Nov./Dec.).

[91] F. Seidel, H. Staffeldt, "Frequency and Angular Resolution for Measuring, Presenting, and Predicting Loudspeaker Polar Data", presented at the 100th Convention of the Audio Engineering Society, *J. Audio Eng. Soc. (Abstracts)*, convention paper 4209 (1996 Jul./Aug.).

[92] EASE software, http://ease.afmg.eu.

[93] ULYSSES software, http://www.ifbcon.de/software/e.php.

[94] CADP2 software, www.jblpro.com/pub/technote/ssdm_ch8b.pdf.

[95] CLF Group, http://www.clfgroup.org/.

[96] W. Ahnert, S. Feistel, J. Baird, P. Meyer, "Accurate Electroacoustic Prediction Utilizing the Complex Frequency Response of Far-Field Polar Measurements", presented at the 108th Convention of the Audio Engineering Society, *J. Audio Eng. Soc. (Abstracts)*, vol. 48, pp. 357, convention paper 5129 (2000 Apr.).

[97] S. Feistel, W. Ahnert, "The Significance of Phase Data for the Acoustic Prediction of Combinations of Sound Sources", presented at the 119th Convention of the Audio Engineering Society, *J. Audio Eng. Soc. (Abstracts)*, convention paper 6632 (2005 Dec.).

[98] M. S. Ureda, "Apparent Apex", presented at the 102nd Convention of the Audio Engineering Society, *J. Audio Eng. Soc. (Abstracts)*, convention paper 4467 (1997 Mar.).

[99] J. Vanderkooy, "The Low-Frequency Acoustic Center: Measurement, Theory, and Application", presented at the 128th Convention of the Audio Engineering Society, *J. Audio Eng. Soc. (Abstracts)*, convention paper 7992 (2010 May).

[100] S. Feistel, W. Ahnert, C. Hughes, B. Olson, "Simulating the Directivity Behavior of Loudspeakers with Crossover Filters", presented at the 123rd Convention of the Audio Engineering Society, *J. Audio Eng. Soc. (Abstracts)*, convention paper 7254 (2007 Oct.).

[101] S. Feistel, A. Goertz, "Digitally Steered Columns: Comparison of Different Products by Measurement and Simulation" presented at the 135th Convention of the Audio Engineering Society, *J. Audio Eng. Soc. (Abstracts)*, convention paper 8935 (2013 Oct.).

[102] EASE SpeakerLab, http://speakerlab.afmg.eu.

[103] C. Heil, "Sound Fields Radiated by Multiple Sound Sources Arrays", presented at the 92th Convention of the Audio Engineering Society, *J. Audio Eng. Soc. (Abstracts)*, convention paper 3269 (1992 March).

[104] M. S. Ureda, "The Convolution Method for Horn Array Directivity Prediction", presented at the 96th Convention of the Audio Engineering Society, *J. Audio Eng. Soc. (Abstracts)*, convention paper 3790 (1994 Feb).

[105] W. Ahnert, "Cluster Design with EASE for Windows", presented at the 106th Convention of the Audio Engineering Society, *J. Audio Eng. Soc. (Abstracts)*, convention paper 4926 (1999 June).

[106] A. Gloukhov, "A Method of Loudspeaker Directivity Prediction Based on Huygens-Fresnel Principle" presented at the 115th Convention of the Audio Engineering Society, *J. Audio Eng. Soc. (Abstracts)*, convention paper 5985 (2003 Dec.).

[107] V. Holtmeyer, *Line Array Loudspeaker System Simulation with the Ulysses CAAD Software*, Diploma thesis, http:/ www.ifbsoft.de/publication/documents/ e_Diploma%20Thesis_VH_Simulation%20Of%20Line-Arrays.pdf (2002 Oct.).

[108] D. W. Gunness, W. R. Hoy, "Improved Loudspeaker Array Modeling - Part 2", presented at the 109th Convention of the Audio Engineering Society, *J. Audio Eng. Soc. (Abstracts)*, convention paper 5211 (2000 Nov.).

[109] M. Di Cola, D. Doldi, "Horn's Directivity Related to the Pressure Distribution at Their Mouth", presented at the 109th Convention of the Audio Engineering Society, *J. Audio Eng. Soc. (Abstracts)*, convention paper 5214 (2000 Nov.).

[110] L. D. Landau, E. M. Lifschitz, *Lehrbuch der Theoretischen Physik*, Band II Klassische Feldtheorie, 12., überarbeitete Auflage (Akademie Verlag, Berlin, 1997).

[111] W. Ahnert, R. Feistel, "Prediction of Wave Fields Synthesized by Speaker Cluster - Needs and Limitations", presented at the 100th Convention of the Audio Engineering Society, *J. Audio Eng. Soc. (Abstracts)*, convention paper 4146 (1996 Jul./Aug.): J. Bohlmann, "Fast Anti-Aliasing und Winkel-Abtasttheorem", private communication (Hamburg, 1998).

[112] B. B. Baker, E. T. Copson, *The Mathematical Theory of Huygens Principle*, 2nd ed. (Oxford University Press, London, 1953).

[113] E. Hering, R. Martin, M. Stohrer, *Physik für Ingenieure*, fünfte, überarbeitete Auflage (VDI-Verlag, Düsseldorf, 1995).

[114] E. G. Williams, *Fourier Acoustics*, 1st ed. (Academic Press, London, 1999).

[115] M. Möser (Ed.), *Messtechnik der Akustik*, 1st ed. (Springer, Berlin, 2009).

[116] Martin Audio, Omniline, www.omniline-ma.com.

[117] Kling & Freitag, SEQUENZA 10N, http://www.kling-freitag.de.

[118] A. Thompson, "Automated Splay Angle Calculation for Line Array Loudspeaker Systems", presented at the 22nd Reproduced Sound, *Proceedings of the Institute of Acoustics*, vol. 28, no. 8 (2006).

[119] S. Feistel, W. Ahnert, "The Effect of Sample Variation Among Cabinets of a Line Array on Simulation Accuracy" presented at the 127th Convention of the Audio Engineering Society, *J. Audio Eng. Soc. (Abstracts)*, convention paper 7842 (2009 Oct.).

[120] H. Margenau, G. M. Murphy, *Die Mathematik für Physik und Chemie*, Band II, (Verlag Harri Deutsch, Frankfurt a. M., 1967).

[121] K. Binder, D. W. Heermann, *Monte Carlo Simulation in Statistical Physics*, 4th ed. (Springer, Berlin, 2002).

[122] W. H. Press, S. A. Teukolsky, W. T. Vetterling, B. P. Flannery, *Numerical Recipes in C*, 2nd ed. (Cambridge University Press, Cambridge, 1997).

[123] Renkus-Heinz, Iconyx IC-8, http://www.renkus-heinz.com.

[124] Electro-Voice, XLC, http://www.electro-voice.com.

[125] R. L. Stratonovich, *Topics in the Theory of Random Noise*, Vol. 1, English ed. (Gordon and Breach, New York, 1963).

[126] S. Feistel, M. Sempf, K. Köhler, H. Schmalle, "Adapting Loudspeaker Array Radiation to the Venue Using Numerical Optimization of FIR Filters" presented at the 135th Convention of the Audio Engineering Society, *J. Audio Eng. Soc. (Abstracts)*, convention paper 8937 (2013 Oct.).

[127] M. Urban, C. Heil, C. Pignon, C. Combet, P. Bauman, "The Distributed Edge Dipole (DED) Model for Cabinet Diffraction Effects", *J. Audio Eng. Soc.*, vol. 52, pp. 1043-1059 (2004 Oct.).

[128] K. Attenborough, S. Taherzadeh, H. E. Bass, X. Di, R. Raspet, G. R. Becker, A. Güdesen, A. Chrestman, G. A. Daigle, A. LEsprance, Y. Gabillet, K. E. Gilbert, Y. L. Li, M. J. White, P. Naz, J. M. Noble, H. A. J. M. van Hoof, "Benchmark cases for outdoor sound propagation models", *J. Acoust. Soc. Am.*, vol. 97, pp. 173-191 (1995).

[129] S. N. Vecherin, D. K. Wilson, V. E. Ostashev, "Incorporating source directionality into outdoor sound propagation calculations", *J. Acoust. Soc. Am.*, vol. 130, pp. 3608-3622 (2011).

[130] L. D. Landau, E. M. Lifschitz, *Lehrbuch der Theoretischen Physik*, Band VI Hydrodynamik, 5., überarbeitete Auflage (Akademie Verlag, Berlin, 1991).

[131] G. A. Daigle, J. E. Piercy, T. F. W. Embleton, "Effects of atmospheric turbulence on the interference of sound waves near a hard boundary", *J. Acoust. Soc. Am.*, vol. 64, pp. 622-630 (1978).

[132] G. A. Daigle, "Correlation of the phase and amplitude fluctuations between direct and ground-reflected sound", *J. Acoust. Soc. Am.*, vol. 68, pp. 297-302 (1980).

[133] G. A. Daigle, J. E. Piercy, T. F. W. Embleton, "Line-of-sight propagation through atmospheric turbulence near the ground", *J. Acoust. Soc. Am.*, vol. 74, pp. 1505-1513 (1983).

[134] S. F. Clifford, R. J. Lataitis, "Turbulence effects on acoustic wave propagation over a smooth surface", *J. Acoust. Soc. Am.*, vol. 73, pp. 1545-1550 (1983).

[135] E. M. Salomons, "A coherent line source in a turbulent atmosphere", *J. Acoust. Soc. Am.*, vol. 105, pp. 652-657 (1999).

[136] E. M. Salomons, V. E. Ostashev, S. F. Clifford, R. J. Lataitis, "Sound propagation in a turbulent atmosphere near the ground: An approach based on the spectral representation of refractive-index fluctuations", *J. Acoust. Soc. Am.*, vol. 109, pp. 1881-1893 (2001).

[137] V. E. Ostashev, E. M. Salomons, S. F. Clifford, R. J. Lataitis, D. K. Wilson, Ph. Blanc-Benon, D. Juvé, "Sound propagation in a turbulent atmosphere near the ground. A parabolic-equation approach", *J. Acoust. Soc. Am.*, vol. 109, pp. 1894-1908 (2001).

[138] L. D. Landau, E. M. Lifschitz, *Lehrbuch der Theoretischen Physik*, Band IX Statistische Physik, 4., überarbeitete Auflage (Akademie Verlag, Berlin, 1992).

[139] G. Vojta, M. Vojta, *Taschenbuch der Statistischen Physik*, (Teubner, Stuttgart, 2000).

[140] L. Schimansky-Geier, T. Pöschel (Eds.), *Stochastic Dynamics*, (Springer, Berlin, 1997).

[141] Thies Clima, Adolf Thies GmbH & Co. KG, http://www.thiesclima.com.

[142] Gill Instruments, Gill Instruments Ltd., http://www.gill.co.uk.

[143] EASERA, http://easera.afmg.eu.

[144] JCGM 100:2008, "Guide to the Expression of Uncertainty in Measurement (GUM)", Geneva, Switzerland, http://www.bipm.org/en/publications/guides/gum.html.

Bibliography

[145] W. Blumen, R. Banta, S. P. Burns. D. C. Fritts, R. Newsom, G. S. Poulos, J. Sun, "Turbulence statistics of a Kelvin-Helmholtz billow event observed in the night-time boundary layer during the Cooperative Atmosphere-Surface Exchange Study field program", *Dynamics of Atmospheres and Oceans*, vol. 34, pp. 189-204 (2001).

[146] M. A. Kallistratova, "Acoustic Waves in the Turbulent Atmosphere: A Review", Journal of Atmospheric and Oceanic Technology, vol. 19, pp. 1139-1150 (2002).

[147] T. Foken, "Turbulenter Energieaustausch zwischen Atmosphäre und Unterlage", *Berichte des Deutschen Wetterdienstes*, Nr. 180 (Selbstverlag des Deutschen Wetterdienstes, Offenbach am Main, 1990).

[148] EASE Focus, http://focus.afmg.eu.

[149] M. Abramowitz, I. A. Stegun, *Handbook of Mathematical Functions*, 9th printing (Dover Publications, New York, 1972).

[150] I. S. Gradshteyn, I. M. Ryzhik, *Table of Integrals, Series and Products*, 6th ed. (Academic Press, San Diego, 2000).

[151] I. N. Bronstein, K. A. Semendjajew, *Taschenbuch der Mathematik*, 10. Auflage (B.G. Teubner, Leipzig, 1969).

Curriculum Vitae

<u>Personal</u>

Date of Birth   August 6th, 1977

Family Status   Married, two children

<u>Education</u>

1990-1996   Secondary School, Gymnasium CJD Rostock

1997-1999   Pre-Diploma (Vordiplom) in Physics,
                         University of Rostock

1999-2004   Master's Degree (Diplom) in Theoretical Physics,
                         Humboldt University Berlin, Thesis: "Bifurcations in
                         Systems of Coupled Phase Oscillators with Noise"

<u>Professional Experience</u>

Since 1996   Co-developer of the electro-acoustic
                         simulation software EASE 3.0 - 4.4

1996-1997   Civil service, care of highly disabled children,
                         Rostock

Since 2000   Co-founder and managing director of
                         SDA Software Design Ahnert GmbH, Berlin

Since 2008   Co-founder and managing director of
                         AFMG Technologies GmbH, Berlin

Stefan Feistel, Spring of 2014

Bisher erschienene Bände der Reihe

Aachener Beiträge zur Technischen Akustik

ISSN 1866-3052

Alle erschienenen Bücher können unter der angegebenen ISBN-Nummer direkt online (http://www.logos-verlag.de) oder per Fax (030 - 42 85 10 92) beim Logos Verlag Berlin bestellt werden.